Treatment of Agroindustrial Biomass Residues

Sílvio Vaz Jr.

Treatment of Agroindustrial Biomass Residues

A Sustainable Approach

Sílvio Vaz Jr.
Embrapa Agroenergia
Parque Estação Biológica
Brasília, DF, Brazil

ISBN 978-3-030-58852-6 ISBN 978-3-030-58850-2 (eBook)
https://doi.org/10.1007/978-3-030-58850-2

This Springer imprint is published by the registered company Springer Nature Switzerland AG
The registered company address is: Gewerbestrasse 11, 6330 Cham, Switzerland

I would like to dedicate this book to my aunt Natália Vaz (in memoriam). She was my second mother in my life.

Preface

The agroindustry based on the plant biomass processing is a global sector that promotes wealth from several crops and their products such as food, feed, fibres, bioenergy, among others. Unfortunately, the high production and productivity of the raw materials (e.g. grains) and their processed products also generate wastes and residues in great amounts, which demands technologies for their treatment in order to turn the agriculture and its production chains more sustainable. From this observation, the subject of this book is present and discusses technologies that allow applied strategies of treatment with positive environmental, societal and economic impacts to the society.

Chapter 1 presents the global biomass production and availability from crops and agrofood processing industry, the main types of plant biomass and its relation with UN Sustainable Development Goals. Nowadays, the world crop production is in the order of 7.26 Gtonnes of total production with a generation of 140 Gtonnes of dry biomass waste. This huge amount of residues creates an environmental problem that demands strategies and technologies for their treatment in order to promote economic value and social development, reducing negative impacts on the environment.

Chapter 2 introduces the basis of sustainability, considering that it is a complex concept which looks to the future of our resources and life quality by means of innovative business strategies, taking into account economic, societal and environmental impacts to be evaluated by metrics, as E-factor and life cycle assessment.

Chapter 3 introduces the basis of environmental chemistry, presenting the renewability of the chemical elements in the environment by means of the biogeochemical cycles and the main environmental matrices. The carbon footprint, a derived application of this cycle, is presented as a strategy to understand and to reduce the generation of CO_2 from agroindustrial activities—considering that it is a greenhouse gas. Organic matter is also discussed in order to understand the biomass transformation in the environment and its direct relationship with sustainability.

Chapter 4 deals with the introduction of more relevant technologies of monitoring and treatment to be applied to biomass residues. To achieve more environmentally and healthy friendly agribusiness is a necessary technology of monitoring —such as real-time monitoring and probes—and technologies of treatment—such as chemical, biochemical, thermochemical and physical processes for the generated residues. Furthermore, the proposal of renewable carbon-based technologies for agroindustrial purposes is presented associated with a proposal of added value to agroindustrial chains based on their residues. Agroindustrial wastes and residues represent one of the main potential sources of supply of raw material for chemistry and other products, due to its large quantity produced and its well-established chains, which facilitates their usages.

Chapter 5 defines the main available technologies and most adequate strategies to treat agroindustrial residues applying chemical, biochemical, thermochemical and physical processes. As the choice of a treatment strategy is not a simple task, in some cases will be necessary an association of processes to achieve the better result. It is associated with aspects of environmental management and circular economy. Moreover, are presented and discussed several examples of treatments of wastes and residues.

Finally, Chap. 6 deals with the more relevant information described by the previous chapters in order to highlight the practical application of them, by means of remarks and conclusions.

The presentation of case studies aims to expand the reader's knowledge and demonstrate examples of application of the technologies proposed by the author.

Good lecture!

Brasília, Brazil Sílvio Vaz Jr.

Acknowledgements The author thanks Dr. Zander Navarro and Dr. Gilmar Souza Santos, scientists at Brazilian Agricultural Research Corporation (Embrapa), for the help with the revision of the text content. Thanks also to Springer team for enabling the transformation of an idea into a book. Finally, thanks to my wife Ana and to my daughter Elena for the patience and attention during the manuscript writing.

Contents

Introduction

1

Abstract

The biomass production and uses come from the first human activities to survive the inhospitable environment. With the agriculture development, its production reached out huge quantities, generating wealth associated with environmental issues, which demands scientific knowledge and technologies of control and treatment to reduce negative impacts by means the modern sustainable vision. Nowadays, the world crop production is in the order of 7.26 Gtonnes of total production with a generation of 140 Gtonnes of dry biomass waste. This huge amount of residues creates an environmental problem that demands strategies and technologies for their treatment in order to promote economic value and social development, reducing negative impacts on the environment. This chapter presents an outlook of the agricultural biomass production and their waste and residues generated with a worldwide distribution and availability.

1.1 Global Biomass Agricultural Production and Availability from Crops and Agrofood Processing Industry

Biomass can be defined as "material produced by the growth of microorganisms, plants or animals" [7], taking into account that the plant biomass is the main subject of this book.

The biomass production and uses come from the first human activities to survive the inhospitable environment. With the agriculture development, its production reached out huge quantities, generating wealth associated with environmental issues, which demands scientific knowledge and technologies of control and treatment to reduce negative impacts by means the modern sustainable vision.

S. Vaz Jr., *Treatment of Agroindustrial Biomass Residues*,
https://doi.org/10.1007/978-3-030-58850-2_1

Agroindustry is the set of activities related to the transformation of raw materials from agriculture, livestock, aquaculture or forestry. The degree of transformation varies widely depending on the objectives of agro-industrial companies. For each of these raw materials, agroindustry is a segment that ranges from the supply of agricultural inputs to the consumer. Compared to other industrial segments of the economy, it has a certain originality due to three fundamental characteristics of raw materials:

- Seasonability
- Perishability
- Heterogeneity

According the World Bank, agricultural development is one of the most powerful tools to end extreme poverty, boost shared prosperity and feed a projected 9.7 billion people by 2050, accounting for one-third of global gross-domestic product (GDP) [19]. Figure 1.1 depicts the main types of products that can be obtained from the agroindustrial advanced biomass processing. We can see a large amount of products obtained by means this industrial processing, with several value chains: materials, chemical inputs for agriculture, energy, materials, and food and animal feed.

We can observe that, for all products highlighted in Fig. 1.1 their processes could generate wastes or residues in the liquid, solid or gaseous state. For instance —but not only:

- Chemical inputs: liquids from reactive medium with potential pollutants
- Energy: gas (CO_2) from combustion process
- Food and animal feed: solid lignocellulosic residues, liquids with high content of organic matter
- Fuels: liquids from reactive medium with potential pollutants
- Materials: solids, as micropollutants (e.g., microplastics)

Figure 1.2 depicts an agroindustry facility dedicated to grains processing.

According Higashikawa et al. [5] the characteristics of plant residues vary depending on the plant species, plant tissues, and soil chemical and physical properties, but are generally nutrient-poor than animal or municipal wastes.

Regarding to the global annual generation of biomass waste, Tripathi et al. [12] estimates it in the order of 140 Gtonnes and it presents significant management problems, as discarded biomass can have negative environmental impacts. Bentsen and Felby [1] determine the potential of biomass residues to four major selected countries, according their amounts of residues in fresh weight:

- China: 716 Mtonnes
- United States of America: 682 Mtonnes
- India: 605 Mtonnes
- Brazil: 451 Mtonnes

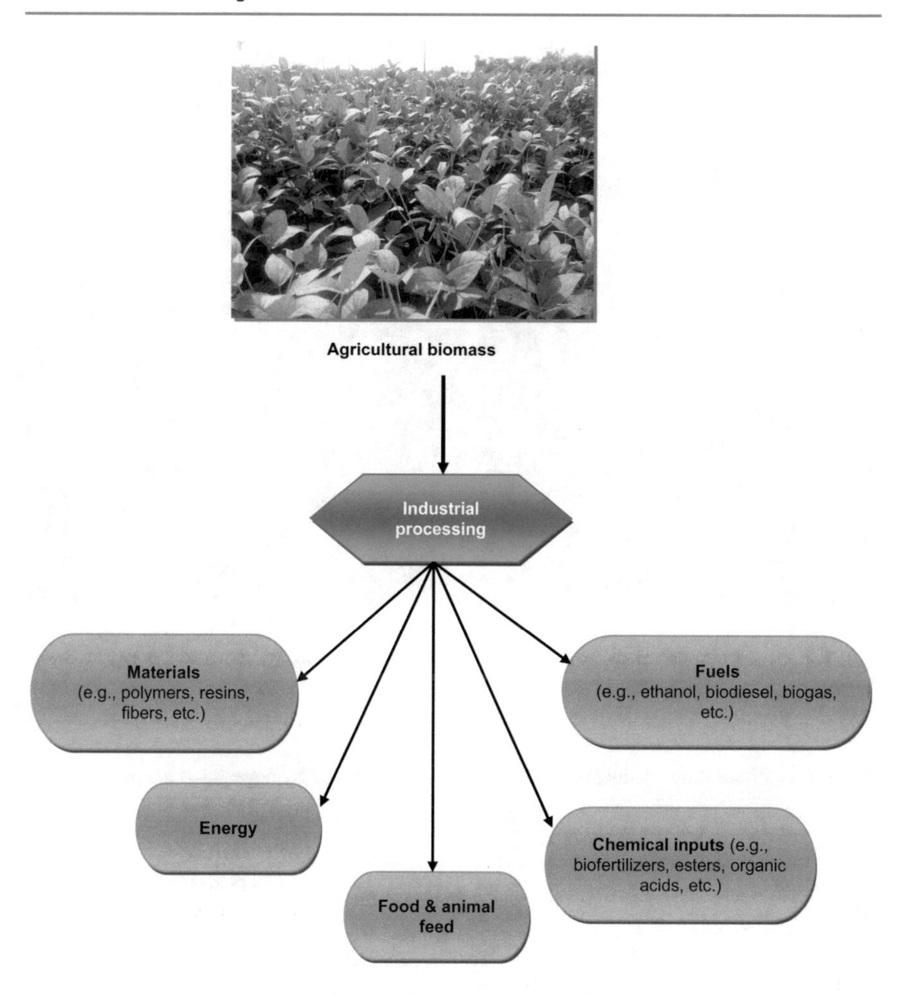

Fig. 1.1 Products that can be obtained from agricultural biomass by means their industrial processing, as physical (e.g., chopping, grinding); chemical (e.g., esterification, combustion); and biochemical (e.g., fermentation). Author

These countries are the main global crops and livestock producers with a well-established agricultural trade. Moreover, the total amount of 2454 Gtonnes of these four countries give us an idea of the huge generation of residues from agroindustrial sources to be treated in order to guarantee absence of negative impacts on the environment and public health. On the other hand, these same residues can be used as raw material for several products, from energy to chemicals.

Table 1.1 describes the world crop production, with 7.26 Gtonnes. Cereals is the main class, with a production of approximately 3 Gtonnes that indicates their relevance for food and animal feed.

Fig. 1.2 An agroindustrial facility dedicated to grains processing (soybean and corn) in Brazil. Author

Table 1.1 World crop production in the year of 2018, divided by aggregated crops

Item	Unit	Value
Cereals	Tonnes	2,962,867,626
Citrus fruit	Tonnes	152,448,800
Coarse grain	Tonnes	1,446,822,305
Fruit	Tonnes	867,774,832
Pulses	Tonnes	92,277,859
Roots and tubers	Tonnes	832,131,696
Treenuts	Tonnes	18,399,741
Vegetables	Tonnes	1,088,839,427

Source Food and Agriculture Organization of the United Nations (http://www.fao.org/faostat/en/#data/QD). Reproduced with permission from Food and Agriculture Organization of the United Nations

Table 1.2 describes the world crops processed production with a highlighted position for the beer of barley (180 Gtonnes) followed by sugar raw centrifugal (176 Gtonnes).

The agroindustry has the objective of transforming the raw materials of livestock, aquaculture, forestry and agriculture, in order to prolong its availability and value. This is because the agroindustry consists of a physical environment equipped for the preparation and transformation of agricultural raw materials.

Table 1.2 World crops processed production in the year of 2014

Item	Unit	Value
Beer of barley	Tonnes	180,332,523
Cotton lint	Tonnes	26,156,675
Cottonseed	Tonnes	46,988,046
Margarine, short	Tonnes	14,199,002
Molasses	Tonnes	60,965,075
Oil, coconut (copra)	Tonnes	3,106,474
Oil, cottonseed	Tonnes	5,036,141
Oil, groundnut	Tonnes	5,031,479
Oil, linseed	Tonnes	686,498
Oil, maize	Tonnes	3,189,137
Oil, olive, virgin	Tonnes	3,050,390
Oil, palm	Tonnes	57,328,872
Oil, palm kernel	Tonnes	6,602,838
Oil, rapeseed	Tonnes	25,944,831
Oil, safflower	Tonnes	100,751
Oil, sesame	Tonnes	1,634,327
Oil, soybean	Tonnes	45,704,551
Oil, sunflower	Tonnes	15,848,036
Palm kernels	Tonnes	15,329,986
Sugar raw centrifugal	Tonnes	176,938,569
Wine	Tonnes	29,105,841

Source Food and Agriculture Organization of the United Nations (http://www.fao.org/faostat/en/#data/QD). Reproduced with permission from Food and Agriculture Organization of the United Nations

In the agroindustry, everything is set up in order to add value to the end-product, e.g., food, maintaining its original characteristics or improve them, as an increasing in shelf life without reducing food quality. Then, agroindustry can be summarized as the process of industrialization of products in the agricultural sector.

Table 1.3 describes the world crops processed production according regions. In Africa beer of barley highlights with 14.8 Mtonnes; in Northern America beer of barley highlights with 24.5 Mtonnes; in Central America sugar raw centrifugal highlights with 11.6 Mtonnes; in South America sugar raw centrifugal highlights with 45.5 Mtonnes; in Asia beer of barley highlights with 64 Mtonnes. This large table give us a brief idea of the residue generation and availability. However, it can vary according factors as weather, economy, epidemic and pandemic, among others not described here.

The huge global crop production generates also a huge production of waste biomass, by means their tillage systems and processing.

Table 1.4 describes the world crop residues, highlighting those generated from nutrients application—wheat is the main source of this residue with 9.9 Gtonnes, followed by corn or maize with 9.3 Gtonnes.

Table 1.3 World crops processed production according regions in the year of 2014

Area	Item	Unit	Value
Area	Item	Unit	Value
Africa	Beer of barley	Tonnes	13,815,581
Africa	Cotton lint	Tonnes	1,636,724
Africa	Cottonseed	Tonnes	2,741,307
Africa	Margarine, short	Tonnes	377,274
Africa	Molasses	Tonnes	4,381,155
Africa	Oil, coconut (copra)	Tonnes	108,602
Africa	Oil, cottonseed	Tonnes	332,643
Africa	Oil, groundnut	Tonnes	1,199,137
Africa	Oil, linseed	Tonnes	37,825
Africa	Oil, maize	Tonnes	151,352
Africa	Oil, olive, virgin	Tonnes	404,800
Africa	Oil, palm	Tonnes	2,305,503
Africa	Oil, palm kernel	Tonnes	317,860
Africa	Oil, rapeseed	Tonnes	78,270
Africa	Oil, safflower	Tonnes	1190
Africa	Oil, sesame	Tonnes	683,027
Africa	Oil, soybean	Tonnes	591,705
Africa	Oil, sunflower	Tonnes	635,790
Africa	Palm kernels	Tonnes	843,105
Africa	Sugar raw centrifugal	Tonnes	11,350,889
Africa	Wine	Tonnes	1,272,933
Northern America	Beer of barley	Tonnes	24,494,400
Northern America	Cotton lint	Tonnes	3,593,000
Northern America	Cottonseed	Tonnes	4,649,320
Northern America	Margarine, short	Tonnes	3,966,400
Northern America	Molasses	Tonnes	2,198,000
Northern America	Oil, cottonseed	Tonnes	277,000
Northern America	Oil, groundnut	Tonnes	95,000
Northern America	Oil, linseed	Tonnes	113,600
Northern America	Oil, maize	Tonnes	1,877,300
Northern America	Oil, olive, virgin	Tonnes	10,700
Northern America	Oil, rapeseed	Tonnes	3,820,100
Northern America	Oil, safflower	Tonnes	33,674
Northern America	Oil, soybean	Tonnes	9,987,600
Northern America	Oil, sunflower	Tonnes	166,400
Northern America	Sugar raw centrifugal	Tonnes	7,762,000
Northern America	Wine	Tonnes	3,354,663
Central America	Beer of barley	Tonnes	9,448,663
Central America	Cotton lint	Tonnes	305,405

(continued)

Table 1.3 (continued)

Area	Item	Unit	Value
Central America	Cottonseed	Tonnes	478,703
Central America	Margarine, short	Tonnes	152,861
Central America	Molasses	Tonnes	3,848,000
Central America	Oil, coconut (copra)	Tonnes	128,761
Central America	Oil, cottonseed	Tonnes	37,742
Central America	Oil, groundnut	Tonnes	23,239
Central America	Oil, linseed	Tonnes	2142
Central America	Oil, maize	Tonnes	29,464
Central America	Oil, olive, virgin	Tonnes	1900
Central America	Oil, palm	Tonnes	1,327,274
Central America	Oil, palm kernel	Tonnes	122,474
Central America	Oil, rapeseed	Tonnes	602,200
Central America	Oil, safflower	Tonnes	18,600
Central America	Oil, sesame	Tonnes	18,287
Central America	Oil, soybean	Tonnes	738,161
Central America	Oil, sunflower	Tonnes	10,400
Central America	Palm kernels	Tonnes	304,005
Central America	Sugar raw centrifugal	Tonnes	11,609,236
Central America	Wine	Tonnes	39,519
Caribbean	Beer of barley	Tonnes	879,358
Caribbean	Cotton lint	Tonnes	392
Caribbean	Cottonseed	Tonnes	704
Caribbean	Margarine, short	Tonnes	16,829
Caribbean	Molasses	Tonnes	732,905
Caribbean	Oil, coconut (copra)	Tonnes	22,885
Caribbean	Oil, cottonseed	Tonnes	155
Caribbean	Oil, groundnut	Tonnes	2352
Caribbean	Oil, linseed	Tonnes	14
Caribbean	Oil, maize	Tonnes	4
Caribbean	Oil, palm	Tonnes	47,000
Caribbean	Oil, palm kernel	Tonnes	6500
Caribbean	Oil, soybean	Tonnes	20,835
Caribbean	Palm kernels	Tonnes	13,000
Caribbean	Sugar raw centrifugal	Tonnes	2,385,745
Caribbean	Wine	Tonnes	12,080
South America	Beer of barley	Tonnes	23,073,982
South America	Cotton lint	Tonnes	1,840,090
South America	Cottonseed	Tonnes	3,369,619
South America	Margarine, short	Tonnes	911,993
South America	Molasses	Tonnes	17,024,369

(continued)

Table 1.3 (continued)

Area	Item	Unit	Value
South America	Oil, coconut (copra)	Tonnes	21,582
South America	Oil, cottonseed	Tonnes	358,297
South America	Oil, groundnut	Tonnes	106,263
South America	Oil, linseed	Tonnes	8710
South America	Oil, maize	Tonnes	191,497
South America	Oil, olive, virgin	Tonnes	45,676
South America	Oil, palm	Tonnes	1,931,687
South America	Oil, palm kernel	Tonnes	279,275
South America	Oil, rapeseed	Tonnes	125,485
South America	Oil, safflower	Tonnes	808
South America	Oil, sesame	Tonnes	6851
South America	Oil, soybean	Tonnes	15,868,306
South America	Oil, sunflower	Tonnes	1,157,992
South America	Palm kernels	Tonnes	619,538
South America	Sugar raw centrifugal	Tonnes	45,467,486
South America	Wine	Tonnes	3,154,902
Asia	Beer of barley	Tonnes	64,049,548
Asia	Cotton lint	Tonnes	17,512,739
Asia	Cottonseed	Tonnes	33,935,583
Asia	Margarine, short	Tonnes	4,911,067
Asia	Molasses	Tonnes	25,383,419

Source Food and Agriculture Organization of the United Nations (http://www.fao.org/faostat/en/#data/QD). Reproduced with permission from Food and Agriculture Organization of the United Nations

Gas emissions (CO_2 and N_2O) and nutrients application (e.g., N-fertilizers) are the main sources of residues generated by tillage. Unfortunately, this compilation did not present the amount of lignocelullosic residues, as it is the main source of plant biomass residues (as seen ahead).

1.2 Main Types of Plant Biomass for Agroindustrial Processing

Based on the high heterogeneity and a consequent large chemical complexity, plant biomass become it the raw material for various ending products such as energy, food, chemicals, and materials (Fig. 1.1). According the large availability, we can consider four types of plant biomass for (agro)industrial purposes, which serve as raw material for those products, with a great economic interest and to which we turn our attention: oil crop or oleaginous, saccharides (or sugary), starch and lignocellulosic. Soybean (*Glycine max*) and palm oil (*Elaeis guinensis*) are examples of oil

Table 1.4 World crop residues of the year 2017

Element	Item	Unit	Value
Residues	Barley	kg of nutrients	1,821,137,638
Emissions (N_2O)	Barley	Gigagrams	35
Emissions (CO_2eq)	Barley	Gigagrams	10,867.6
Residues	Beans, dry	kg of nutrients	431,367,614
Emissions (N_2O)	Beans, dry	Gigagrams	8.3
Emissions (CO_2eq)	Beans, dry	Gigagrams	2574
Residues	Maize	kg of nutrients	9,286,749,653.8
Emissions (N_2O)	Maize	Gigagrams	178.8
Emissions (CO_2eq)	Maize	Gigagrams	55,418.7
Residues	Millet	kg of nutrients	344,942,463.2
Emissions (N_2O)	Millet	Gigagrams	6.6
Emissions (CO_2eq)	Millet	Gigagrams	2058.4
Residues	Oats	kg of nutrients	291,636,160.7
Emissions (N_2O)	Oats	Gigagrams	5.6
Emissions (CO_2eq)	Oats	Gigagrams	1740.3
Residues	Potatoes	kg of nutrients	801,639,371.3
Emissions (N_2O)	Potatoes	Gigagrams	15.4
Emissions (CO_2eq)	Potatoes	Gigagrams	4783.8
Residues	Rice, paddy	kg of nutrients	9,155,085,230.1
Emissions (N_2O)	Rice, paddy	Gigagrams	176.2
Emissions (CO_2eq)	Rice, paddy	Gigagrams	54,631
Residues	Rye	kg of nutrients	149,673,281.4
Emissions (N_2O)	Rye	Gigagrams	2.9
Emissions (CO_2eq)	Rye	Gigagrams	893.2
Residues	Sorghum	kg of nutrients	821,650,763.1
Emissions (N_2O)	Sorghum	Gigagrams	15.8
Emissions (CO_2eq)	Sorghum	Gigagrams	4903.2
Residues	Soybeans	kg of nutrients	4,523,479,877.7
Emissions (N_2O)	Soybeans	Gigagrams	87.1
Emissions (CO_2eq)	Soybeans	Gigagrams	26,993.9
Residues	Wheat	kg of nutrients	9,895,078,429.6
Emissions (N_2O)	Wheat	Gigagrams	190.5
Emissions (CO_2eq)	Wheat	Gigagrams	59,048.9

Source Food and Agriculture Organization of the United Nations (http://www.fao.org/faostat/en/#data/QD). Reproduced with permission from Food and Agriculture Organization of the United Nations

plants species; sugarcane (*Saccharum* spp.) and sorghum (*Sorghum bicolor* (L.) Moench) are saccharides; maize (*Zea mays*) is a starchy biomass; bagasse, straw and wood biomass are lignocellulosic biomass [16]. Figure 1.3 shows the classification of the sources of plant biomass.

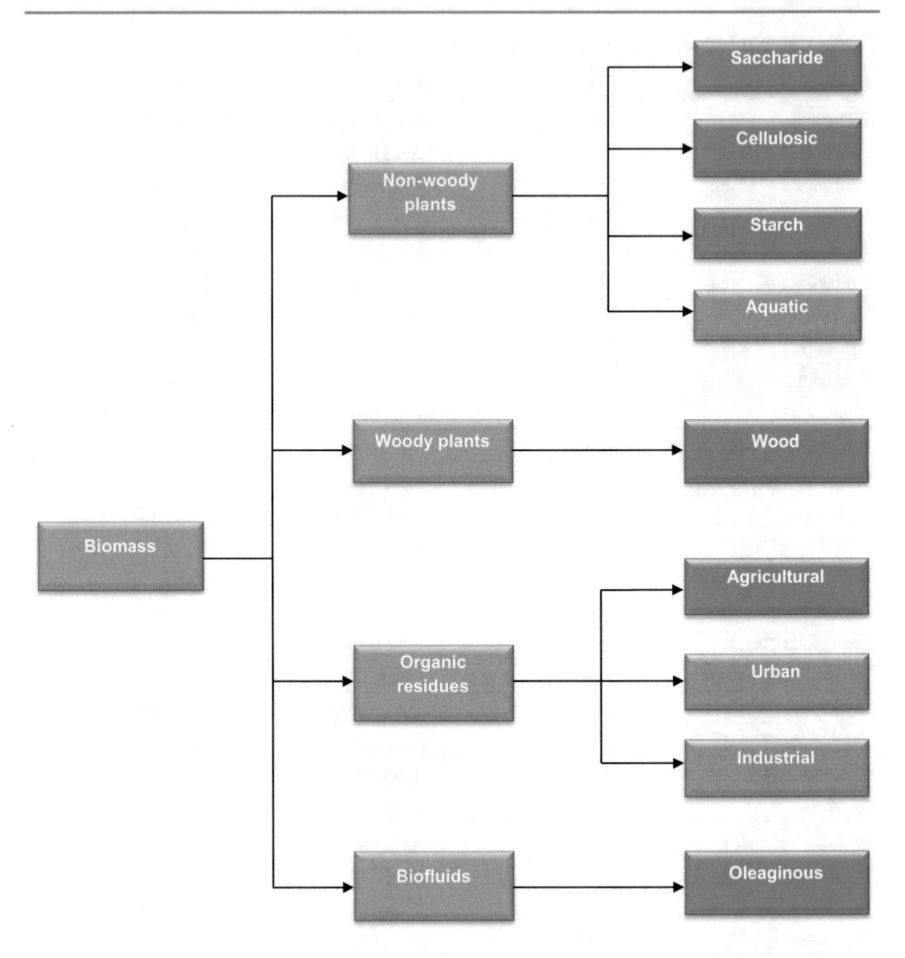

Fig. 1.3 Sources of biomass; gray boxes represent the most used biomass types for industrial and research & development & innovation (R&D&I) activities [15]. Reproduced with permission from the Royal Society of Chemistry

Figures 1.4, 1.5, 1.6, 1.7, 1.8 and 1.9 shows the chemical structure of components of these biomass types, and Tables 1.5, 1.6, 1.7 and 1.8 shows their wt./wt. percentages.

Oleaginous biomass is one that has superior fatty acids and their esters stored inside seeds or grains, with fatty acids with chains of different sizes and number of unsaturations (lipids).

Oleaginous plants are used as feedstock for the follow agroindustrial products:

- Cooking oil obtained from extraction
- Raw material for the agrochemical and chemical industries
- Raw material for biodiesel

H_3C COOH

Stearic acid

H_3C COOH

Oleic acid

H_3C COOH

Linoleic acid

Fig. 1.4 Some chemical structures of fatty acids from oleaginous plants such as soybean. Author

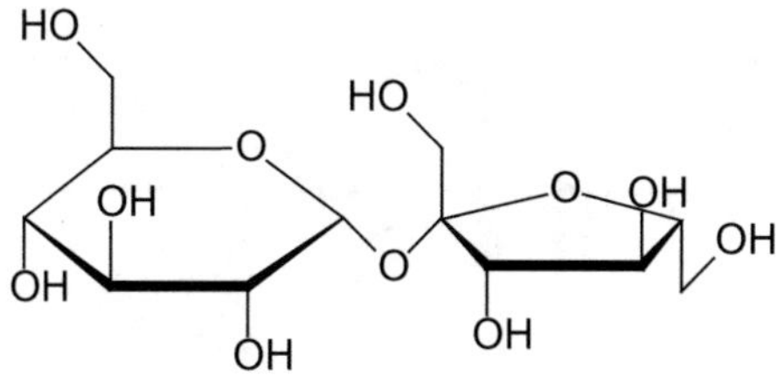

HO HO O O OH OH O OH OH OH OH

Fig. 1.5 Chemical structure of sucrose, a disaccharide present in sugarcane. The D-glucose moiety is on the left and the D-fructose moiety is on the right linked by α-β-D-disaccharide bonds. Author

O CH_2O O OH OH O CH_2O O OH OH O

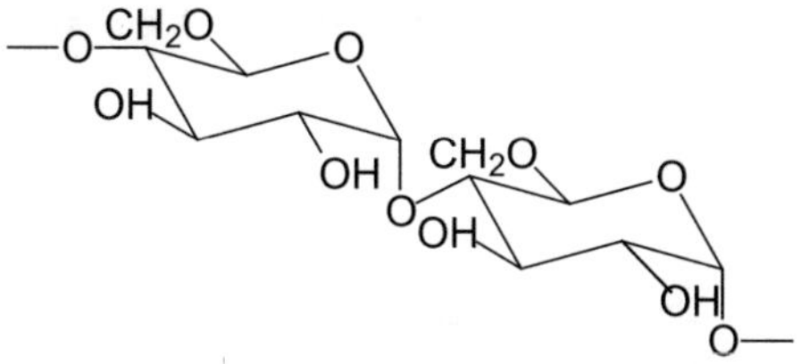

Fig. 1.6 Chemical structure of starch polymer; the glucose unities (monomers) are linked by α-1-4-D-disaccharide bonds. Author

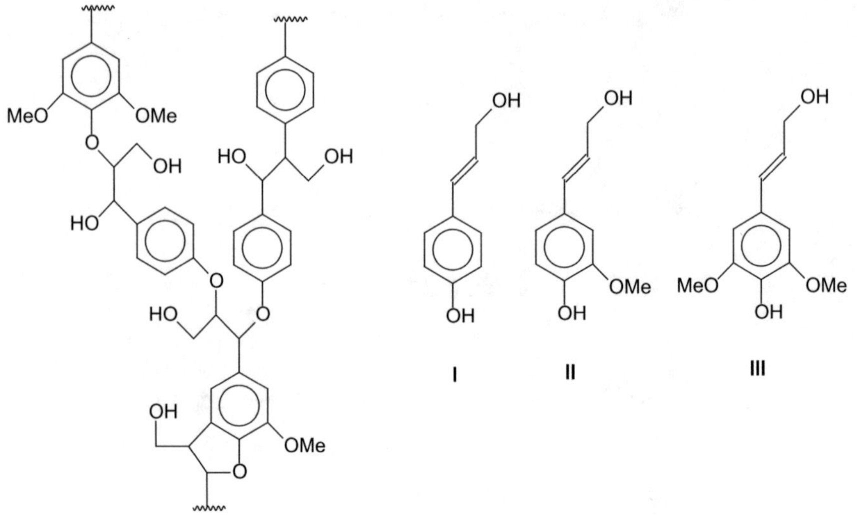

Fig. 1.7 Lignin structure (left) and its precursors (right): (I) *p*-coumaryl alcohol, (II) coniferyl alcohol, and (III) sinapyl alcohol. Author

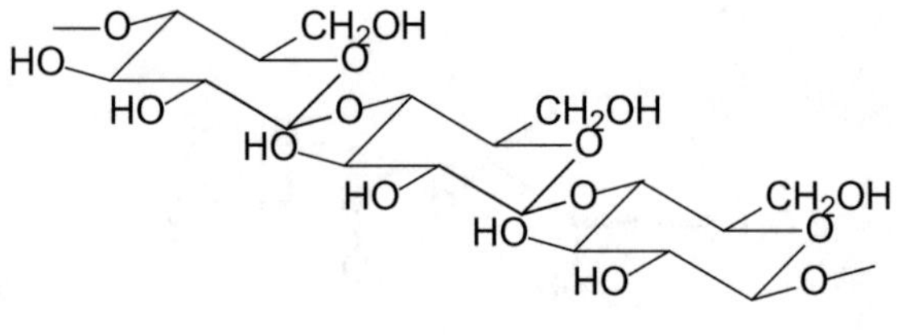

Fig. 1.8 Chemical structure of cellulose; the glucose units are linked by 1,4-β-D bond. Author

Saccharide biomass is one in which the main source of sugar is sucrose, which is a disaccharide formed by glucose and fructose, the latter being its constituent monosaccharides, which are hexoses (C6).

Saccharides or sugary plants are used as feedstock for the follow agroindustrial products:

- Sugar (saccharose)
- Ethanol
- Raw material for the chemical industries
- Raw material for renewable polymers

Starch biomass has this name because its main chemical constituent is starch, which is a polymer—polysaccharide—whose monomer is glucose, which is a hexose, or six-carbon sugar (C6), also called monosaccharide.

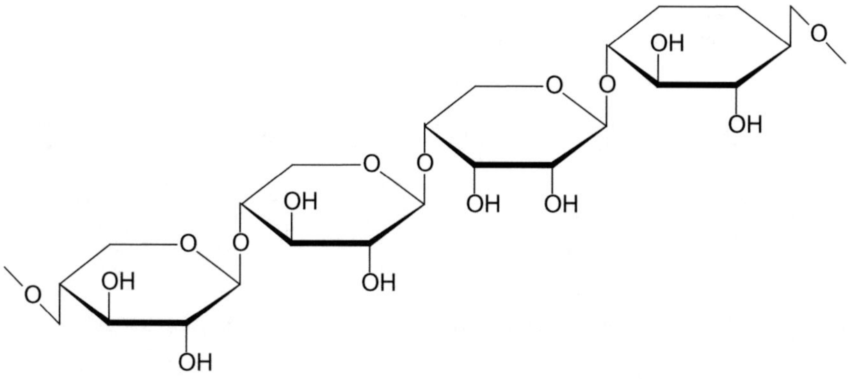

Fig. 1.9 Chemical structure of hemicellulose; the oligomeric units composed of D-glucose and pentoses (mainly D-xylose) are linked by means of a 1,4-β-D bond. Author

Table 1.5 Fatty acid composition (wt./wt., %) of oils extracted from oleaginous biomass [3]. Reproduced with permission from Elsevier

FAC (%)	Oil type			
	Peanut oil	Rapeseed oil	Soybean oil	Linseed oil
C16:0	10.80 ± 0.01	4.04 ± 0.01	10.14 ± 0.01	5.76 ± 0.01
C18:0	4.18 ± 0.02	1.76 ± 0.00	4.32 ± 0.00	4.22 ± 0.01
C18:1 ω9	41.64 ± 0.02	61.44 ± 0.07	22.62 ± 0.00	18.61 ± 0.01
C18:2 ω6	35.78 ± 0.01	19.30 ± 0.01	52.16 ± 0.04	16.07 ± 0.01
C18:3 ω3	–	7.63 ± 0.01	6.59 ± 0.01	50.61 ± 0.01
C20:0	1.62 ± 0.03	0.61 ± 0.00	0.41 ± 0.01	0.19 ± 0.00
C20:1 ω7	1.03 ± 0.01	1.37 ± 0.02	0.38 ± 0.01	0.12 ± 0.02
C22:0	2.91 ± 0.01	–	–	0.11 ± 0.01
C24:0	1.42 ± 0.00	–	–	0.08 ± 0.00

Results are means ± SD of fatty acid composition. *FAC* fatty acid composition

Table 1.6 Chemical composition of broth extracted from sugarcane [4] and sweet sorghum [9]. Reproduced with permission from Elsevier

Plant	wt./wt. m/m sucrose	wt./wt. m/m glucose	wt./wt. m/m organic acid
Sugarcane	85.3	–	24
Sweet sorghum	14.8	1.5	–

Table 1.7 Chemical composition of corn grain flour [11], cassava [2], and potato [8]. Reproduced with permission from Elsevier

Plant	% wt./wt. starch	% wt./wt. protein	% wt./wt. fiber	% wt./wt. others
Corn (flour of grain)	90.1	6.5	0.52	1.99 (lipd)
Cassava (pulp)	83.8	1.5	2.5	0.2 (lipd)
Potato (pulp)	71.5	8.6	5.4	–

Table 1.8 Chemical composition of cellulosic biomasses [14]. Reproduced with permission from Elsevier

Biomass	% wt./wt. cellulose	% wt./wt. hemicellulose	% wt./wt. lignin
Barley straw	48.6	29.7	21.7
Corn cobs	48.1	37.2	14.7
Grasses	34.2	44.7	21.1
Sugarcane bagasse	42.7	33.1	24.2
Rice husks	43.8	31.6	24.6
Wheat straw	44.5	33.2	22.3
Eucalyptus	52.7	15.4	31.9

As sugary plants, starchy plants are used as feedstock for the follow agroindustrial products:

- Sugar (glucose)
- Ethanol
- Raw material for the chemical industries
- Raw material for renewable polymers

Lignocellulosic biomass is the most abundant compared to the others already treated, as it is formed by cellulose, hemicellulose and lignin, which are the three components of the cell wall and the morphological structure of plants—cellulose and hemicellulose are polysaccharide polymers and lignin a phenolic macromolecule.

Lignocellulosic plants are used as feedstock for the follow agroindustrial products:

- Energy by combustion (co-generation)
- Raw material for renewable polymers
- Cellulose and their derivatives
- Lignin and their derivatives
- Sugars from cellulose (glucose) and hemicellulose (xylose)

It is the main source of agricultural residues due to the fact that all plants are constituted by lignin, cellulose and hemicellulose.

1.3 Biomass as Renewable Feedstock According the UN Sustainable Development Goals

Currently, agriculture must constantly become increasingly more sustainable, with the reduction in its negative impacts on the environment being matched with the demand to increase its positive impacts on society and economy. These are challenges and, at same time, opportunities for new production and processing systems.

Regarding to the modern approaches for the processing industry, *green chemistry* emerges in the 1990s in countries such as the United States and England, spreading very fast to the world, as a new philosophy in academia and industry to break old paradigms of chemistry such as large waste generation and intensive use of petrochemicals through a holistic view of processes in laboratories and industries (Anastas and Kirchhoff 2002). This approach, described in 12 principles proposes to consider, among other aspects, the reduction of waste generation, the atomic and energy economy, and the use of renewable raw materials (Anastas and Warner 1998). In the case of plant biomass, the seventh principle—*use of renewable raw materials* - stands out as a great strategic opportunity for segments related to chemical industry worldwide. An example of market segments that may be positively impacted by green chemistry principles and the use of biomass are [17]:

- Polymers and materials for various applications
- Chemical commodities such as monomers
- Pharmaceuticals, cosmetics and hygiene products
- Fine chemicals (agrochemicals, catalysts, etc.) and specialties
- Fuels and energy.

The United Nations [13] established 17 Sustainable Development Goals to promote sustainable global growth (Fig. 1.10). Goal 2 (*zero hunger*) is closely related to agriculture and food security; according to this goal, "*a profound change in the global food and agriculture system is needed if we are to nourish the 815 million people who are hungry today and the additional 2 billion people expected to be undernourished by 2050.*" Thus, agriculture has a paramount responsibility to find ways to provide food for such increasing demand in the years ahead. At the same time, devising ways to reduce impacts associated with agricultural production that could be considered harmful to the environment is also key.

The global biomass supply comprises 11.4 Gtonnes year^{-1} of dry matter [6]; however, the quantity and quality depends on the tillage system and the agroindustrial production in each country.

According to the Organization for Economic Co-operation and Development [10], *bioeconomy* "refers to the set of economic activities relating to the invention, development, production and use of biological products and processes". From this definition, key aspects surrounding the sustainability—to be seen in the Chap. 2—of bioeconomy development are:

Fig. 1.10 The 17 UN sustainable development goals. *Source* United Nations [13]. Reproduced with permission from United Nations (https://www.un.org/sustainabledevelopment/)

- The use of biomass as feedstock for future production;
- The design and building of biorefineries for the manufacture of a range of fuels, chemicals and materials, and also for electricity generation;
- The use of biotechnologies such as synthetic biology, metabolic engineering and gene editing.

Of course, this concept possesses a wide scope and it's applicability to agriculture is obvious. Moreover, bioeconomy is aligned with the UN Sustainable Development Goals due to the promotion of sustainable value chains based on renewable resources.

New opportunities for bioeconomy are largely found in manufacturing, biochemistry and agriculture, but strategies also need to include accelerated innovations for food security and resource protection [18].

In general, a strategy to explore agriculture and their wastes and residues, according to the bioeconomy, could take into account these topics:

- Bioproducts and biorefineries: refers to the supply of products resulting from the conversion of biomass into bioproducts (e.g., biofuels, materials and chemicals).
- Biomass chemistry and technology: refers to the supply of biomass on a renewable basis and the development of processes based on the use of biomass.
- Production and use of biomass: refer to the most efficient use of available biomass.

- Renewable energy: refers to the supply of energy-related products from renewable energy sources.
- Climate change: considers more promising alternatives or strategies of the reduction of global warming and adaptation to changes.
- Food and nutritional security: consider regular and permanent access to quality food without compromising access to other needs.
- Use and exploitation of natural resources: consider obtaining benefits from the use of natural resources.
- Valuation of natural resources and ecosystem services: considers environmental benefits resulting from human interventions in the dynamics of ecosystems.
- Transversal to the bioeconomy: presents issues related to investment, the regulatory framework and market and is considered important for the development and application of the concept.

Then, bioeconomy is the most adequate approach to the applied for the use of biomass as an industrial renewable feedstock.

Parts of Sect. 1.3 are reproduced with permission from Brazilian Agricultural Research Corporation (Embrapa).

1.4 Conclusions

Agricultural development is one of the most powerful tools to end extreme poverty, boost shared prosperity and feed a projected 9.7 billion people by 2050, and the world produces 7.26 Gtonnes of it for agroindustrial processing.

However, this same biomass production generates 140 Gtonnes of waste with a heterogeneous chemical composition in different physical states that need the best technical and economic approaches to reduce their impact on the environment.

This huge amount of residues can generates opportunities for its use as renewable industrial feedstock, according green chemistry and bioeconomy concepts, in a close relationship with the UN Sustainable Development Goals.

References

1. Bentsen NS, Felby C (2010) Technical potentials of biomass for energy services from current agriculture and forestry in selected countries in Europe, The Americas and Asia. Forest & Landscape Working Papers No. 54, 31 pp. Forest & Landscape Denmark, Frederiksberg
2. Charles AL, Sriroth K, Huang T-C (2005) Proximate composition, mineral contents, hydrogen cyanide and phytic acid of 5 cassava genotypes. Food Chem 92:615–620
3. Chen J, Zhang L, Li Q, Wang M, Dong Y, Yu X (2020) Comparative study on the evolution of polar compound composition of four common vegetable oils during different oxidation processes. LWT—Food Sci Technol 129:109538
4. Faria S, Petkowicz CLO, De Morais SAL, Terrones MGH, De Resende MM, De França FP, Cardoso VL (2011) Characterization of xanthan gum produced from sugar cane broth. Carbohydrate Polym 86:469–476

5. Higashikawa FS, Silva CA, Bettiol W (2010) Chemical and physical properties of organic residues. Revista Brasileira de Ciência do Solo 34:1743–1752
6. International Energy Agency. IEA Bioenergy (2017) The role of biomass, bioenergy and biorefining in a circular economy. https://www.iea-bioenergy.task42-biorefineries.com/upload_mm/9/1/0/64005b9b-e395-497e-b56f-8c145fdfc18d_D5%20The%20role%20of%20Biomass%20Bioenergy%20and%20Biorefining%20in%20a%20Circular%20Economy%20-%20Paris%20meeting%20-%20version%20170105.pdf/. Accessed Apr 2020
7. International Union of Pure and Applied Chemistry (2020) Gold book. https://goldbook.iupac.org/terms/view/B00660. Accessed Apr 2020
8. Liu Q, Tarn R, Lynch D, Skjodt NM (2007) Physicochemical properties of dry matter and starch from potatoes grown in Canada. Food Chem 105:897–907
9. Mamma D, Chistakopoulus P, Koullas D, Kekos D, Macris BJ, Koukios E (1995) An alternative approach to the bioconversion of sweet sorghum carbohydrates to ethanol. Biomass Bioenergy 8:99–103
10. Organization for Economic Co-operation and Development (2018) Meeting policy challenges for a sustainable bioeconomy. http://www.oecd.org/sti/policy-challenges-facing-a-sustainable-bioeconomy-9789264292345-en.htm. Accessed Apr 2020
11. Sandhu KS, Singh N, Malhi NS (2007) Some properties of corn grains and their flours I: Physicochemical, functional and chapatti-making property of flours. Food Chem 101:938–946
12. Tripathi N, Hills CD, Singh RS, Atkinson CJ (2019) Biomass waste utilization in low-carbon products: harnessing a major potential resource. Nature NPJ Clim Atmos Sci 35:1
13. United Nations (2020) Sustainable development goals. https://www.un.org/sustainabledevelopment/sustainable-development-goals/. Accessed Apr 2020
14. Vassilev SV, Baxter D, Andersen LK, Vassileva CG, Morgan TJ (2012) An overview of the organic and inorganic phase composition of biomass. Fuel 94:1–33
15. Vaz S Jr (2014) Analytical techniques for the chemical analysis of plant biomass and biomass products. Anal Methods 6:8094–8105
16. Vaz S Jr (ed) (2016) Analytical techniques and methods for biomass. Springer Nature, Cham
17. Vaz S Jr (ed) (2018) Biomass and green chemistry—building a renewable pathway. Springer Nature, Cham
18. von Braun J (2018) Bioeconomy—the global trend and its implications for sustainability and food security. Glob Food Security 19:81
19. World Bank (2020) Agriculture and food home. https://www.worldbank.org/en/topic/agriculture/overview#1. Accessed Apr 2020

Basis of Sustainability for Biomass

2

Abstract

As a theme that call very attention in the twenty-first century by the global society, sustainability is a complex concept. However, sustainability looks to the future of our resources and life quality by means innovative business strategies, take into account economic, societal and environmental impacts to be evaluated by metrics, as E-factor and life cycle assessment. This Chapter deals with these definitions and application to agroindustrial biomass and its residues.

2.1 Definition of Sustainability

As a theme that call very attention in the twenty-first century by the global society, sustainability is a complex concept. The development of sustainable processes and products is a continuous searching in a very technological world that can become a certain good more valuable. Probably the most suitable definition comes from the UN World Commission on Environment and Development: "*sustainable development is development that meets the needs of the present without compromising the ability of future generations to meet their own needs.*" That is, sustainability looks to the future of our resources and life quality by means smart strategies, which involves the 17 Sustainable Development Goals (SDGs) [14]. These SDGs are depicted in the Fig. 2.1.

Goals 6 (clean water and sanitation), 7 (affordable and clean energy), 9 (industry, innovation and infrastructure), 12 (responsable consumption and production) and 13 (climate action) are closely related to the biomass processing and the treatment of its waste and residues in order to achieve sustainable productive chains.

S. Vaz Jr., *Treatment of Agroindustrial Biomass Residues*,
https://doi.org/10.1007/978-3-030-58850-2_2

Fig. 2.1 The 17 Sustainable Development Goals (SDGs) [14]. Reproduced with permission from United Nations (https://www.un.org/sustainabledevelopment/)

The sustainability concept comprises three components or pillars:

- Economic impacts
- Social impacts
- Environmental impacts.

These impacts can be *positives* or *negatives* according the appropriate metrics and are considerated in details in the next item.

2.2 Economic, Social and Environmental Impacts

The three pillars or components of sustainability are by the time composed by a variety of internal indicators to be considered, as:

- Economic impacts: natural resources use, environmental management, and pollution prevention applied to air, water, land and waste;
- Social impacts: standard of living, education, community, equal opportunities;
- Economic impacts: profit, cost savings, economic growth, research & development.

Several indicators can be listed according the productive chain, turning them suitable to each case of study, i.e., they attends to the demand. Furthermore, Sikdar [12] suggested consider the intersection—or interdependencies—of the pillars with more internal indicators, what can generates:

- Environmental-economic: energy efficiency, subsides/incentives for use of natural resources;
- Social-environmental: environmental laws, natural resources stewardship;
- Economic-social: business ethics, fair trade, worker's rights.

Then, it is expected a positive impact from all three sustainability components in order to deliver more friendly consumer goods to the society.

2.3 Relationship Between Sustainability Components and Biomass Processing

When we consider the agroindustrial biomass processing there are some categories of specific sustainability indicators, mainly for the environmental and social components in order to achieve a reduced environmental footprint and increased societal value [15].

For a reduced environmental footprint:

- Greenhouse gas emissions—*reduction or absence*
- Persistent toxic emissions, as dioxins and furans—*reduction or absence*
- Material intensity—*reduction*
- Ecological impacts—*reduction or absence*
- Land use—*reduction*
- Water intensity—*reduction or absence*
- Energy intensity—*reduction.*

Regarding greenhouse gas emissions, it is related to carbon dioxide and methane released on the atmosphere. Persistent toxic emissions, for instance, release of dioxins and furans (gaseous effluent). For material intensity, the amount of feedstock used in the processing; moreover, for land use carbon dioxide released by tillage. Furthermore, we can consider the treatment of biomass waste and residues as a deployment of the ecological impacts.

For an increased societal value:

- Poverty alleviation—*improvement*
- Health and safety improving—*improvement*
- Asset recovery—*improvement*
- Prosperity and economic resilience—*improvement*
- Biodiversity and ecological resilience—*improvement*

- Resource conservation—*improvement*
- Human dignity—*improvement.*

To achieve the goals for reduced environmental footprint and increased societal value, is paramount the investment in research & development & innovation (R&D&I) to create new "green" technologies, especially for production systems and for transformation processes. These new technologies involves the reduction in water, energy and inputs use allied to cutting-edge scientific assets from areas as nanotechnology and biotechnology.

Furthermore, the concept of circular economy and its application can contribute to a holistic vision of the economic context of the biomass residues harnessing. The European Parliament [3] defines it as "*a model of production and consumption, which involves sharing, leasing, reusing, repairing, refurbishing and recycling existing materials and products as long as possible. In this way, the* ***life cycle of products is extended.*** *In practice, it implies* ***reducing waste*** *to a minimum. When a product reaches the end of its life, its materials are kept within the economy wherever possible. These can be productively used again and again, thereby* ***creating further value***." As an smart strategy for biomass residues we can propose:

- A reduction in the waste generation, by means the use of all by-products and/or coproducts;
- The creation of new value chains from residues harnessing.

2.4 Metrics for Biomass Waste and Residues

The determination and application of a certain metric of sustainability should take into account the industrial ecosystem and its relationship with external factors, as environmental laws, society concerns, and energy and feedstock availability. On this way, Sheldon [11] suggests the use as a metric for biomass conversion to sustainable fuels and molecules the *E-factor*, that take both the carbon dioxide derived from energy consumption and water usage into account; eventually, it can be applied together with life cycle assessment (seen ahead).

The E-factor can be calculated as follows:

$$\text{E - factor} = \text{kg waste/kg product} \tag{2.1}$$

Nevertheless, Sheldon observes a limitation of this E-factor due that it takes only the mass of waste generated into account, when the environmental footprint of waste is determined not only by its amount but also by its nature. However, even with this limitation the E-factor is a simple and practical metric for industrial purposes.

2.5 Life Cycle Assessment

Life cycle is the set of all the necessary steps for a product fulfills its function in the chain of production and it is closely related to industrial ecology—to be seen in the Chap. 3.

The life cycle assessment (LCA) is a very useful tool for the sustainability determination of industrial processes. It involves the evaluation of products and processes within defined domains, e.g., *cradle-to-gate*, *cradle-to-grave, cradle-to-cradle* and *gate-to-gate*, on the basis of quantifiable environmental impact indicators, such as energy usage, greenhouse gas emissions, ozone depletion, acidification, eutrophication, smog formation, and ecotoxicity, in addition to waste generated [8, 9]. A LCA methodology can involve:

- Manufacturing
- Shipping
- Installation
- Indoor air quality
- Performance
- Resource recovery
- Raw materials acquisition.

LCA steps are internationally standardized by the Society of Environmental Toxicology and Chemistry [13] and the International Organization of Standardization [6, 7]. From this standardization, are considered two main components:

- Life cycle inventory: defines the product's representative system; identifies and quantifies interactions of these operations with the environment (exchanges, inputs and outputs of matter and energy between the environment and the system); and handles the collected data.
- Impact assessment: identifies impact categories for each of the inventoried elements; quantifies the contributions of the elements (use of models); and totals for each category the individual contributions.

As a practical example, Bartocci et al. [2] developed a LCA methodology to assess the environmental and economic benefit of the substitution of energy crops (like corn silage) with food waste in anaerobic digestion. Figure 2.2 shows their applied methodology.

To this assessment were considered the follow boundaries:

- Logistic optimization
- Biomass plant mass and energy balances
- Economic index
- Environmental evaluation (e.g., carbon footprint)
- Feasibility tool development.

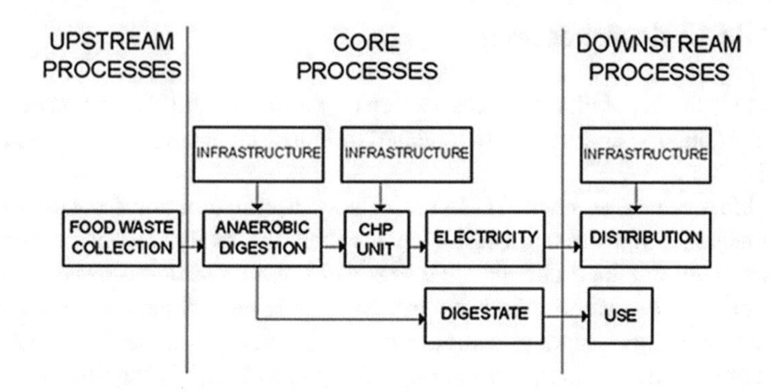

Fig. 2.2 A LCA methodology applied to food waste for energy use; CNP = combined heat & power. *Source* Bartocci et al. [2]. Reproduced with permission from Elsevier

Moreover, were considered as categories of impacts:

- Climate change
- Human toxicity
- Water use
- Natural resources depletion.

Is important consider that LCA contributes to the SDG 13 (climate action).

Nowadays, there are many LCA softwares commercially available as SimaPro[1] and GaBi.[2] As free access Open LCA software can de cited.[3]

2.5.1 An Inventory of Agroindustrial Residues

As previous commented, the inventory is a determinant part of the LCA methodology, defining the product's representative system; and identifying and quantifying interactions of these operations with the environment (exchanges, inputs and outputs of matter and energy between the environment and the system). This information are handling by mean a set of representative collected data.

Porfiro and De Amorim [10] established an inventory of residues from a sugarcane plant, which produces sugar and ethanol. It is presented in the Table 2.1.

From the Table 2.1, it is possible observe that there are several necessities for treatment (e.g., waste oil, fusel oil, contaminated tissue and empty pesticide packing) and later utilization or reutilization (bagasse, vinasse, boiler ashes and filter cake) of those residues in a sugarcane plant which can contribute to promote the sustainability of sugar and ethanol productions.

[1]https://simapro.com/.
[2]http://www.gabi-software.com/.
[3]http://www.openlca.org/.

Table 2.1 Inventory of residues from a sugarcane plant in Brazil

Residue	Amount generated	Destination	Time to retreat	Class[a]
Bagasse	800 ton day^{-1}	Burning in the boiler	Stocked	No inert
Vinasse	14 L: 1L ethanol	Fertirrigation	Daily	No inert
Boiler ashes	14,000 kg day^{-1}	Tillage	Stocked	Inert
Filter cake	3.52 ton $hour^{-1}$	Tillage	Constantly collected	No inert
Fusel oil	17 L $hour^{-1}$	Commercialization	Collected every 2 months	Hazardous
Sugarcane washing water	2.5 m^3 ton^{-1}	Re-use	Returns to the process	Inert
Waste oil	2.4 m^3 $month^{-1}$	Sale for recycling	Collected every semester	Hazardous
Used tires	20 unities $month^{-1}$	Sale for recycling	Collected every semester	Inert
Contaminated tissue	300 unities $month^{-1}$	Exchange for a new	Collected every semester	Hazardous
Individual protection equipment	96 unities $month^{-1}$	Exchange for a news	Collected every 5 months	No inert
Metallic scrap	10 ton $month^{-1}$	Sale for recycling	Collected every semester	No inert
Empty pesticide packing	800 unities $month^{-1}$	Returned to the supplier	Collected every 4 months	Hazardous

[a]Classification according the Brazilian norm ABNT NBR 10004 [1]
Source adapted from Porfiro and De Amorim [10]. Reproduced with permission from Amorim (author)

2.6 Carbon Footprint

The carbon footprint, which means the complete understanding of CO_2 generation and fate, is a very useful tool to estimate the sustainability of processes and products, mainly the environmental component.

In this way, García et al. [5] studied the carbon footprint for sugar production and determined that:

- Carbon footprint for sugar ranges from 0.45 to 0.63 kg CO_2e kg^{-1}.[4]
- Agricultural stage has the largest uncertainty in the estimation of emissions.
- Nitrogen fertilizers contributes the most with carbon emissions.
- There are multiple steps to reduce the carbon footprint of sugar.

[4]According to the Food and Agriculture Organization of the United Nations [4], emissions are measured in CO_2 equivalent (CO_2 eq)—a metric used to compare different greenhouse gases.

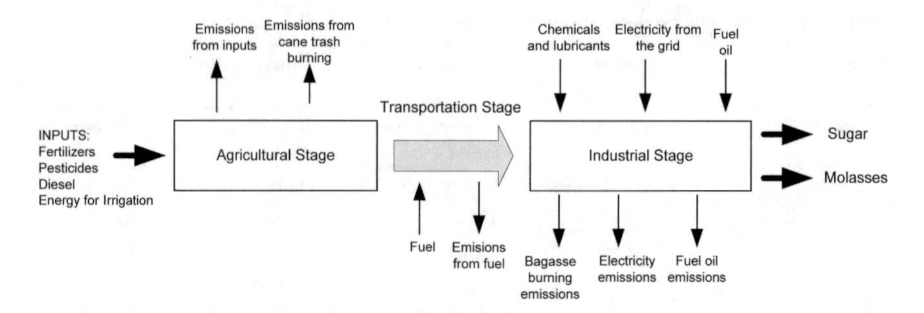

Fig. 2.3 System boundaries, inputs, outputs and sources of greenhouse gas emissions used to calculate carbon footprint values. *Source* García et al. [5]. Reproduced with permission from Elsevier

- Carbon footprint for sugar production could be reduced by implementing efficient cogeneration.

The parameters utilized in this study are depicted in the Fig. 2.3.

2.7 Conclusions

Despite sustainability be a theme that call very attention in the twenty-first century by the global society, it is a complex concept. However, sustainability looks to the future of our resources and life quality by means smart strategies, take into account economic, social and environmental impacts.

To achieve the goals for reduced environmental footprint and increased societal value, is paramount the investment in research & development & innovation to create new "green" technologies, especially for production systems and for transformation processes.

Metrics developed to evaluate these impacts are E-factor and LCA, which can be useful and complementary tools for the agroindustry. For instance, LCA can be applied to food waste to generate energy as an example of treatment of biomass residues according the circular economy approach.

Then, in a practical way sustainability can contribute to more environmental, economic and social friendly processes and products to be deliver to the society.

References

1. Associação Brasileira de Normas Técnicas (2004) ABNT NBR 10004:2004 Resíduos sólidos – classificação (Solid waste—classification). ABNT, São Paulo
2. Bartocci P, Zampilli M, Liberti F, Pistolesi V, Massoli S, Bidini G, Fantozzi F (2020) LCA analysis of food waste co-digestion. Sci Total Environ 709:136187

3. European Parliament (2018) Circular economy: definition, importance and benefits. https://www.europarl.europa.eu/news/en/headlines/economy/20151201STO05603/circular-economy-definition-importance-and-benefits. Accessed April 2020
4. Food and Agriculture Organization of the United Nations (2014) Greenhouse gas emission from agriculture, forestry and other land use. http://www.fao.org/resources/infographics/infographics-details/en/c/218650/. Accessed April 2020
5. García CA, García-Treviño ES, Aguilar-Rivera N, Armendáriz C (2016) Carbon footprint of sugar production in Mexico. J Cleaner Prod 112:2632–2641
6. International Organization of Standardization (2006) ISO 14040:2006 Environmental management—life cycle assessment—principles and framework. ISO, Geneva
7. International Organization of Standardization (2012) ISO 14045: 2012 Environmental management—eco-efficiency assessment of product systems—principles, requirements and guidelines. ISO, Geneva
8. Jiménez-González C, Curzons AD, Constable DJC, Cunningham VL (2004) Cradle-to-gate life cycle inventory and assessment of pharmaceutical compounds. Int J Life Cycle Assessment 9:114–121
9. Monteiro JGM-S, OdeQF Araujo, de Medeiros JL (2009) Sustainability metrics for eco-technologies assessment, part I: preliminary screening. Clean Technol Environ Policy 11:209–214
10. Porfiro SA, De Amorim FR (2012) Inventário de resíduos sólidos em uma usina sucroalcooleira do sudoeste goiano (Inventory of solid residues in a sugarcane plant in southwest Goiás). http://www.unirv.edu.br/conteudos/fckfiles/files/INVENTARIO%20DE%20RESIDUOS%20SOLIDOS%20EM%20UMA%20USINA%20SUCROALCOOLEIRA%20DO%20SUDOESTE%20GOIANO.pdf. Accessed April 2020
11. Sheldon RA (2011) Utilisation of biomass for sustainable fuels and chemicals: molecules, methods and metrics. Catal Today 167:3–13
12. Sikdar SK (2003) Sustainable development and sustainability metrics. AIChE J 49:1928–1932
13. Society of Environmental Toxicology and Chemistry (2020) Life cycle assessment interest group. https://www.setac.org/page/IGLCA. Accessed April 2020
14. United Nations (2020) Sustainable development knowledge platform. https://sustainabledevelopment.un.org/. Accessed April 2020
15. United States Environmental Protection Agency (2012) A framework for sustainability indicators at EPA. https://www.epa.gov/sites/production/files/2014-10/documents/framework-for-sustainability-indicators-at-epa.pdf. Accessed April 2020

Basis of Environmental Chemistry for Biomass

3

Abstract

Chemical elements are renewed in the environment, being removed and returned to nature continuously by means biological, chemical and geological processes, constituting the biogeochemical cycles. The carbon footprint, a derived application of this cycle, became a strategy to understand and to reduce the generation of CO_2 from agroindustrial activities—considering that it is a greenhouse gas. Organic matter originates from the decomposition of residues from plant biomass and animal remains that, through chemical, physical and biological processes, undergo structural modification giving rise to a series of organic compounds, whose main representatives are humic substances. On the other hand, environmental issues can be generated from biomass production and processing as greenhouse gas emissions and water pollution. Then, this Chapter presents concepts of environmental chemistry applied to the context of biomass processing (Parts of this chapter were reproduced with permission from: Vaz Jr. (2019) Scenarios, Challenges and Opportunities for Sustainable Agricultural Chemistry. ISSN 2177-4439, Embrapa Agroenergia, Brasília. Copyright information: 2019, Embrapa Agroenergia, Parts of this chapter were reproduced with permission from: Vaz Jr. (2018) Analytical Chemistry Applied to Emerging Pollutants. Springer Nature, Cham. Copyright information: 2018, Springer Nature,, Parts of this chapter were reproduced with permission from: Vaz Jr. (Ed) (2019) Sustainable Agrochemistry—A Compendium of Technologies. Springer Nature, Cham. Copyright information: 2019, Springer Nature).

S. Vaz Jr., *Treatment of Agroindustrial Biomass Residues*,
https://doi.org/10.1007/978-3-030-58850-2_3

3.1 Dynamic of Chemical Elements in the Environment

All chemical elements present in the Periodic Table have its own dynamics in the environment, i.e., its genesis and fate according several biological, physical and chemical processes which involve degradation and generation of new chemical species. However, the main interest of this Chapter is the organic matter derived from biomass and its chemical elements constituents as oxygen, carbon and nitrogen because, according the [11], biomass is the material produced by the growth of microorganisms, plants or animals, i.e., it is originated from agroindustrial biomass among another sources.

But, in order to a better understanding of their dynamic, is important introduce the main environmental matrices (air, soil and water), as follow, because these matrices have a direct contact with biomass from its seeding to its processing. Furthermore, a matrix as soil could receive products generated by treatment processes applied to biomass waste and residues, e.g., a biofertilizer to improve their functionality.

3.1.1 Air

The nitrogen and oxygen gases are the main constituents of the air in volume (v/v), 78.1% and 20.9%, respectively. However, water vapor, carbon dioxide (CO_2) and argon can also be observed in a more representative proportion in relation to the other gases present in much reduced concentrations. Carbon dioxide and methane (CH_4) have attracted a lot of attention because of global warming—mainly the emission of carbon dioxide in the atmosphere via the burning of fossil fuel leading in an increase in temperature due to the absorption of infrared radiation (wavelength between 10^{-6} to 10^{-3} m). Methane from biochemical processes of biomass degradation can also produce this heating, due to the absorption of the same radiation. Then, both gases are known *greenhouse gases*.

Ozone (O_3), as a constituent of the Earth's protective layer against ultraviolet radiation (wavelength between 10^{-8} m and 10^{-6} m), also raises concern, since the decrease in its atmospheric concentration allows a higher incidence of this radiation, which can lead to skin cancer.

We can observe that the presence of biomass particulate matter from sugarcane harvest and processing can pollute the air causing a respiratory hazard (Le Blond 2017).

Air can be divided into *indoor* (i.e., a closed system) and *outdoor* (i.e., an open system). Table 3.1 depicts examples of pollutants for indoor air. As we can observe, some reactants are products from biomass or biomass-derived.

Table 3.1 Possible reaction products in indoor air with potential emission sources and reactants

Reactants	Products	Possible source
α-pinene	Pinene oxide, pinonaldehyde	Wood, wood-based products
Limonene	Limonene oxide, carvone, formaldehyde	Wood, coating systems
Oleic acid	Heptanal, octanal, nonanal, decanal, 2-decenal	Linoleum, eco-lacquers, nitrocellulose-lacquers, alkyd resins
Linolenic acid	2-pentenal, 2-hexenal, 3-hexenal, 2-heptenal, 2,4-heptedienal, 1-penten-3-one	
Linoleic acid	Hexanal, heptanal, 2-heptenal, octanal, 2-octenal, 2-nonenal, 2-decenal, 2,4-nonadienal, 2,4-decadienal	
Hemicelluloses	Furfural, acetic acid	Cork
PHMP	Benzaldehyde, acetone, benzil	UV-cured coatings
HCPK	Benzaldehyde, cyclohexanone, benzil	UV-cured coatings
2-ethyl-hexyl acetate	Acetic acid, 2-ethyl-1-hexanol	Solvent
Zn-2-ethylhexanoate	2-ethyl-1-hexanoic acid	Stabilizers
n-butylacrylate	*n*-butanol	Acrylate coatings

Source Adapted from [23]. Reprinted with permission from Elsevier

3.1.2 Soil

Soil is the most complex environmental matrix due to the chemical constitution of its organic and inorganic components, and their physical states—soil is formed by chemical substances in the solid, liquid and gaseous states. Then, soils have a natural tendency to interact with different chemical species, among them some pollutants (e.g., persistent organic pollutants).

The distribution of organic matter (OM) and the mineral fraction in layers in the soil is as follows:

- Horizon O (surface): OM in decomposition (0.3 m of depth).
- Horizon A: OM accumulated mixed with the mineral fraction (0.6 m of depth).
- Horizon B: clay accumulation, Fe^{3+}, Al^{3+} and low OM content (approximately 1 m of depth).
- Horizon C: materials from rock mother.

It is expected that the higher the concentration of OM, especially of humic acids present in it—*seen ahead*, the greater the retention capacity of metallic cations in soils, especially in the horizon O, which leads to a reduction in the transport of metallic species in the soil, as the humic substances act as strong complexing agents due to the presence of binder sites formed by carboxylic and phenolic groups [5]—so, we can see here the relevance of biomass and its later degradation into humic substances for the environment dynamic. Therefore, a higher concentration, for example, of bivalent metal cations in the samples of horizons O and A of the soil is expected, considering the effect of the presence of silicate compounds in the metal retention, where a greater capacity of cation exchange capacity (CEC) of the soil denotes a higher availability of binding sites for the metal after the exit of cations or protons associated with these silicates, due to the negative surface charge of the latter.

Figure 3.1 depicts the dynamic of soil organic matter (SOM). This SOM originates, for instance, from agroindustrial residues (e.g., lignocellulosic biomass).

The humic substances (humine, humic and fulvic acids) can contribute to the soil health and functionality. SOM and peat are the main active sources of carbon in terms of reversibility of entry and exit of this element in the biogeochemical cycle of nature [12]. Item 3.2 deals in deep the dynamic of organic matter.

Some pollutants can be observed in soil, as:

- Pesticides, from agricultural activities
- Oil derivatives, from industrial, retail and end-use activities
- Toxic metals, released from mining and industrial activities
- Microplastics, from plastic degradation
- Other molecules from several sources not discriminated here

As for air and water, these chemical species can be monitoring in soil for control and treatment according maximum values and methods from agencies as: European Environmental Agency[1] and US-Environmental Protection Agency[2]. The soil quality is directly related to the water quality because the first can act as a vehicle for pollutant transporting to the second.

[1]https://europa.eu/european-union/documents-publications/reports-booklets_en
[2]https://www.epa.gov/agriculture/agriculture-and-soils

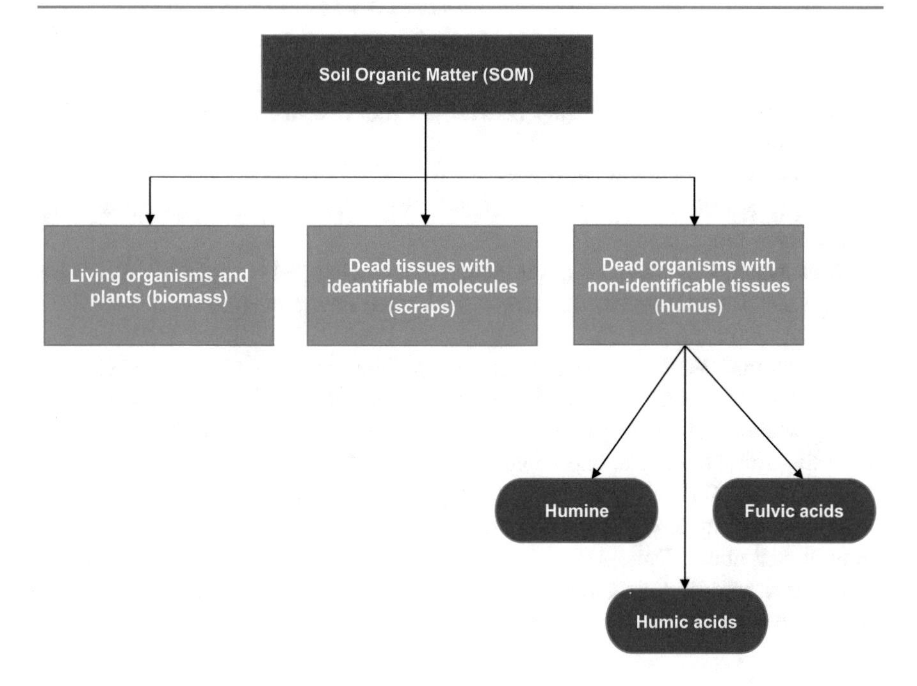

Fig. 3.1 The dynamic of soil organic matter looks its degradation to generate humus, a stock of humic substances. Author

3.1.3 Water

We can consider two distinct types, but which strongly correlate: *surface water* and *groundwater*. The surface water is that found in rivers, lakes, seas and oceans; while groundwater is that found in the aquifers. Drinking-water and wastewater are classifications related to surface and groundwater according their use.

Groundwater, which has a higher concentration of CO_2 gas, is in greater contact with rocks and soil, which leads to a longer dissolution time [18]. It is important to mention that carbonic acid (H_2CO_3) produced by the solubilization of CO_2 when in contact with these materials (rocks and soil) leads to the solubilization of the minerals, releasing their constituent ions. Furthermore, a large variety of suspended material can be found mainly in surface waters and clay, sand and organic matter are examples of particles in suspension. There are also several species of microorganisms present in water, highlighting bacteria such as coliforms and cyanobacteria, which often compromise the quality of water, especially surface water. Water bodies—especially surface water—are the main route of exposition to pollution due to sanitary problems as the absence of wastewater and sewage treatment.

Other water-related matrices are:

- Sediments: naturally occurring material, as rocks, sand and silt, in contact with water bodies, as rivers.
- Sewage or municipal wastewater.
- Sludge: a residual semi-solid material from industrial, water or wastewater treatment processes.
- Wastewater: result of a domestic, industrial, commercial or agricultural activity, with negative impacts to the human health and environment.

We can observe that water is frequently polluted by agroindustrial activities as use of pesticides, uncontrolled use, discharge of effluents from food processing, among others, demanding a wastewater treatment and reuse in agriculture [8], according microbiological and chemical analyses [28].

The World Healh Organization [29] publishes, periodicaly, a guidelines for drinking-water quality control and consideres some chemical species from agricultural activities (Table 3.2). The sources of chemical constituints related to agricultural activities are manures, fertilizers, intensive animal practices and pesticides. Moreover, these species can reach the soil and polluting it.

3.1.4 Biogeochemical Cycles for Oxygen, Carbon and Nitrogen

The chemical elements are renewed in the environment, being removed and returned to nature continuously by means biological, chemical and geological processes, constituting the biogeochemical cycles. Regarding to biomass, we can highlight oxygen, carbon and nitrogen cycles which have a direct relation with the production and processing of agroindustrial biomass.

Oxygen cycle

Oxygen is present in several organic and inorganic molecules, such as water and carbon dioxide. In the atmosphere, it is found in the form of gas (O_2) and is released into the environment through photosynthesis performed by autotrophs organisms. Such gas is consumed by plants and animals through respiration, restarting the cycle. The oxygen and carbon cycle are closely related (Figs. 3.2. and 3.3).

Carbon cycle

Carbon is the fifth most abundant element on the planet and present in organic molecules. In the atmosphere, it is found in the form of carbon dioxide (CO_2), released from the breathing of living beings, decomposition and combustion of organic matter (Fig. 3.6). Thus, CO_2 starts to circulate in the atmosphere and it is removed from the environment through the process of photosynthesis carried out by autotrophs organisms, represented mainly by plants. In the presence of light and

Table 3.2 Guideline values for chemicals from agricultural activities with health significance in drinking-water. Reprinted with permission from World Health Organization

Chemical	Guideline value		Remarks
	µg/l	mg/l	
Non-pesticides			
Nitrate (as NO_3^-)	50 000	50	Based on short-term effects, but protective for long-term effects
Nitrite (as NO_2^-)	3 000	3	Based on short-term effects, but protective for long-term effects
Pesticides used in agriculture			
Alachlor	20[a]	0.02[a]	
Aldicarb	10	0.01	Applies to aldicarb sulfoxide and aldicarb sulfone
Aldrin and dieldrin	0.03	0.000 03	For combined aldrin plus dieldrin
Atrazine and its chloro-s-triazine metabolites	100	0.1	
Carbofuran	7	0.007	
Chlordane	0.2	0.000 2	
Chlorotoluron	30	0.03	
Chlorpyrifos	30	0.03	
Cyanazine	0.6	0.000 6	
2,4-D[b]	30	0.03	Applies to free acid
2,4-DB[c]	90	0.09	
1,2-Dibromo-3-chloropropane	1[a]	0.001[a]	
1,2-Dibromoethane	0.4[a] (P)	0.000 4[a] (P)	
1,2-Dichloropropane	40 (P)	0.04 (P)	
1,3-Dichloropropene	20[a]	0.02[a]	
Dichlorprop	100	0.1	
Dimethoate	6	0.006	
Endrin	0.6	0.000 6	
Fenoprop	9	0.009	
Hydroxyatrazine	200	0.2	Atrazine metabolite
Isoproturon	9	0.009	
Lindane	2	0.002	
Mecoprop	10	0.01	
Methoxychlor	20	0.02	
Metolachlor	10	0.01	
Molinate	6	0.006	
Pendimethalin	20	0.02	
Simazine	2	0.002	
2,4,5-T[d]	9	0.009	
Terbuthylazine	7	0.007	
Trifluralin	20	0.02	

P, provisional guideline value because of uncertainties in the health database

[a] For substances that are considered to be carcinogenic, the guideline value is the concentration in drinking-water associated with an upper-bound excess lifetime cancer risk of 10^{-5} (one additional cancer per 100 000 of the population ingesting drinking-water containing the substance at the guideline value for 70 years). Concentrations associated with estimated upper-bound excess lifetime cancer risks of 10^{-4} and 10^{-6} can be calculated by multiplying and dividing, respectively, the guideline value by 10.

[b] 2,4-Dichlorophenoxyacetic acid.

[c] 2,4-Dichlorophenoxybutyric acid.

[d] 2,4,5-Trichlorophenoxyacetic acid.

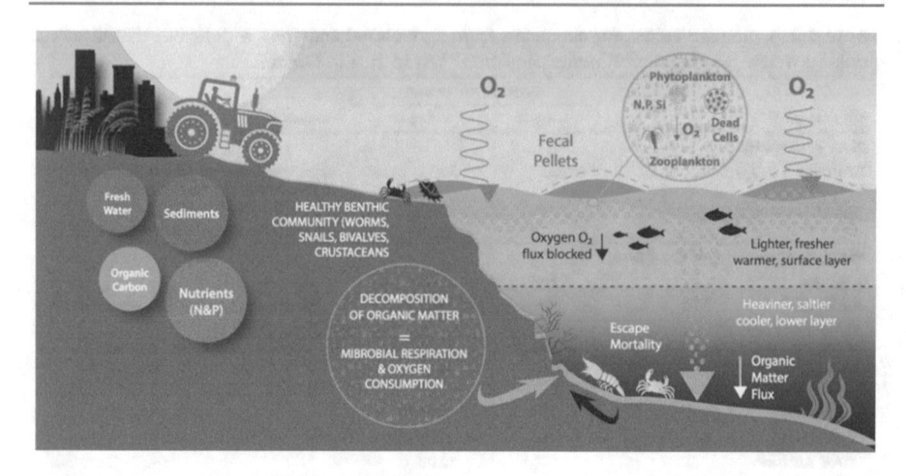

Fig. 3.2 A simplified oxygen biogeochemical cycles in the nature. *Source* https://theanthropocenedashboard.com/category/biogeochemical-flows/oxygen-minimum-zones/

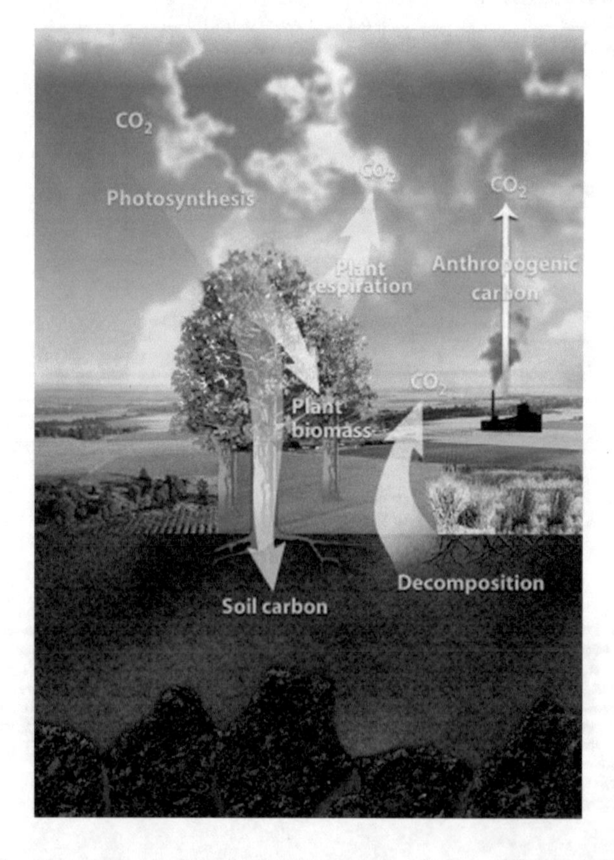

Fig. 3.3 Carbon biogeochemical cycle in the nature *Source* Office of Biological and Environmental Research of the U.S. Department of Energy Office of Science. https://science.energy.gov/ber/

chlorophyll, carbon dioxide and water are converted into glucose ($C_6H_{12}O_6$), with the release of oxygen (O_2), as shown in the Eq. 3.1:

$$6CO_2 + 6H_2O + \text{energy(sunlight)} \rightarrow C_6H_{12}O_6 + 6O_2 \quad (3.1)$$

The *carbon footprint*, a derived application of carbon cycle model, became a strategy to understand and to reduce the generation of CO_2 from agroindustrial activities—considering that it is a greenhouse gas. For instance, Arrieta et al. (2018) obtained the carbon footprints of soybean and maize in Argentina; soybean resulted in 6.06 ton/ton CO_2-eq emitted to the atmosphere, and maize resulted in 5.01 ton/ton CO_2-eq emitted to the atmosphere.[3] This subject is treated in details in Chap. 2.

Nitrogen cycle

About 78% (v/v) of the atmosphere is constituted by nitrogen in the form of gas (N_2). It is a very important element because constitutes proteins and nucleic acids. However, living beings, except for some microorganisms, are unable to incorporate and use nitrogen as a gas, obtaining this nutrient in the form of ammonium ions (NH^{4+}) and nitrate ions (NO_3^-). The nitrogen cycle is divided into four stages (Fig. 3.4) that comprise:

1st stage—Nitrogen fixation

The fixation or assimilation of N_2 can be done through radiation or with the participation of microorganisms, whose process is called *biofixation*. The cyanobacteria of the genera *Nostoc* and *Anabaena* and the bacteria *Azotobacter*, *Clostridium* and *Rhizobium* (live in the roots of legumes) are involved on this biofixation. These organisms fix the amphospheric N_2 in a form usable by living beings, such as ammonium and nitrate. Furthermore, N_2 reacts with hydrogen to form ammonia (NH_3).

2nd stage—Ammonification

Ammonia can be obtained both by the action of biofixers and by the decomposition of proteins, nucleic acids, nitrogenous residues from cadavers and excreted by decomposers (bacteria and fungi).

3rd stage—Nitrification

It is the conversion of ammonia through nitrification that is divided into two parts:

- Nitrosation: converts ammonia (NH_3) into nitrite (NO^{2-}) by the action of nitrifying bacteria (*Nitrosomonas*, *Nitrosococcus*, *Nitrosolobus*) that are autotrophic chemosynthesizers and use the energy of nitrification for the synthesis of organic substances.

[3]According to the [7], emissions are measured in CO_2 equivalent (CO_2 eq)—a metric used to compare different greenhouse gases.

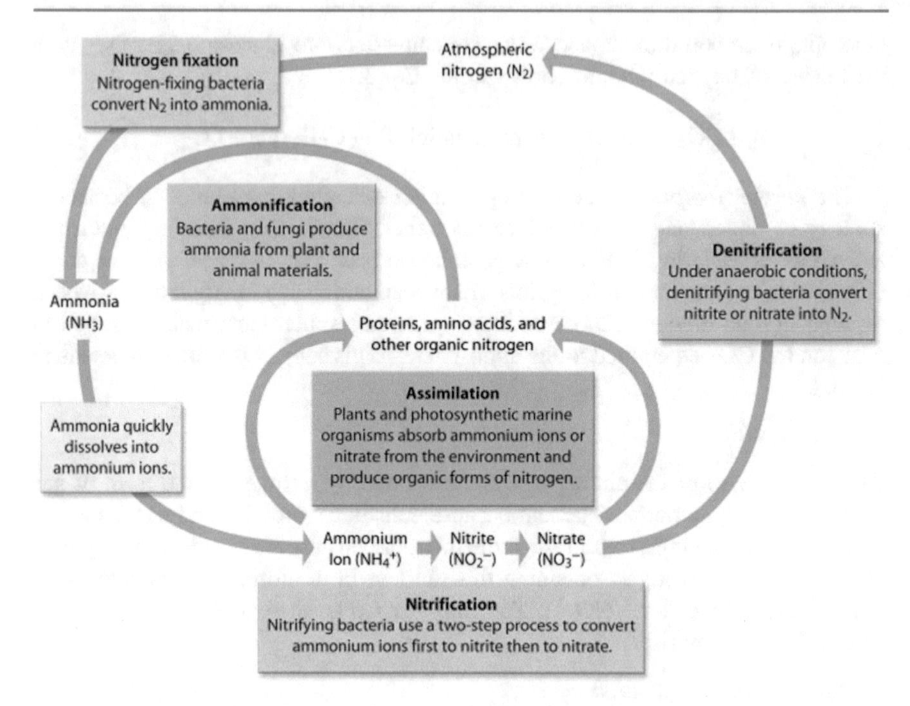

Fig. 3.4 Nitrogen biogeochemical cycle in the nature *Source* Office of Biological and Environmental Research of the U.S. Department of Energy Office of Science. https://science.energy.gov/ber/

– Nitration: the nitrite (NO^{2-}) that is released in the soil is oxidized to nitrate (NO^{3-}) by other autotrophic nitric chemosynthetic bacteria (*Nitrobacter*). Nitrate can be absorbed by vegetables in the manufacture of their proteins and nucleic acids. Nitrogen is released from the body through the excretion of nitrogen products (urea and uric acid) and/or decomposition of dead organisms by the action of decomposers (bacteria and fungi), which also degrade proteins from dead organisms, transforming them into ammonia.

4th stage—Denitrification

Denitrifying bacteria (*Pseudomonas denitrificans*) transform using nitrate to oxidize organic compounds and produce energy, being transformed into gas, restarting the cycle.

3.2 Dynamics of Organic Matter in the Environment

The organic matter (OM) of soil and water originates from the decomposition of residues from plant biomass and animal remains that, through chemical, physical and biological processes, undergo structural modification giving rise to a series of organic compounds, whose main representatives are humic substances (HSs). These HSs contribute fundamentally to the soil properties, as in the case of the sequestration capacity, mobilization and redox effect of organic molecules foreign to the environment (*xenobiotics*) [19], since HSs interact with several organic pollutants, such as pesticides and petroleum derivatives, and inorganics such as toxic metals [5].

HSs are, certainly, the most abundant sources of organic components in nature and are present in all soils and natural waters that contain OM [19]. As seen in item 3.1.2, peat and soil OM (SOM) are the main active sources of carbon in terms of reversibility of entry and exit of this element in the biogeochemical cycle of nature [12].

The constitution of HSs, based operationally on solubility in acid or alkali medium, is as follows [19]:

- Fulvic acids (FAs): fraction soluble in all acidity and alkaline conditions;
- Humic acids (HAs): fraction soluble in alkaline medium;
- Humine: fraction insoluble in any acidity and alkaline conditions.

From these three fractions, HAs are more susceptible to structural changes resulting from soil management practices and the degradation processes of OM [22].

However, the definition of a chemical structure of HAs is a controversial topic, as there are two proposals under discussion: *macromolecular* [21] and *supramolecular* [4, 15], where the macromolecular establishes that the structural variations of HAs are similar those observed in biological macromolecules, such as proteins, polysaccharides, nucleic acids and lignins and depend on the pH and ionic strength of the medium; supramolecular, on the other hand, establishes that HAs in solution forms molecular aggregates that are stabilized by weak interactions, such as hydrogen bonding and hydrophobic interactions (van der Waals). Some studies point out that the supramolecular association of humic components of low molecular weight is the most consistent structural model, based on chromatographic information of the constituent elements of such humic components, such as nitrogen, and of the interactions suffered due to the chemical environment [20], as in the case of the appearance of hydrophobic protective regions under acidic conditions, detected by the decrease in the line width of the electronic paramagnetic resonance spectroscopy (EPR) signal of the semi-quinone radical of the HAs [13]. Furhtermore, [16] presented the strategy of *Humeome* for the HAs fractionation which strongly suggest the supramolecular structure as the most suitable model.

HAs, as biomass degradation products, are rich in phenolic, quinone and aromatic structures (originating mainly from lignin), carboxylic and N-heterocyclic rings (originating from amino acids and proteins), in addition to polysaccharide residues (originating from cellulose and hemicellulose) [6], giving polar and non-polar terminations to the molecular structure.

3.3 Additional Environmental Issues Related to Biomass Production and Processing

CO_2 emission and climate change

The reduction of carbon dioxide emission from agricultural activities is an issue related to warming and climate changes because agricultural systems emit this gas from the land-use [1, 17]. Then, a low-carbon agriculture is a requirement for the twenty-first century.

According the [9], the land-use conversion and soil cultivation have been an important emission source of greenhouse gases (GHGs), as CO_2, to the atmosphere; it is estimated that they are still responsible for about one-third of GHGs emissions. The Intergovernmental Panel on Climate Change (IPCC) [10], estimates that agriculture, forestry and other land use respond by 24% of the global GHGs emissions by economic sector—here, we can consider cultivation of crops and livestock and a possible association with deforestation; and losing only to the electricity and heat production (25%). However, agricultural ecosystems can capture CO_2 from atmosphere by sequestering carbon in biomass, dead organic matter, and soils, which offset approximately 20% of emissions from this sector [26].

A strategy for a low-carbon agriculture can comprise [2]:

- Recovery of degraded pastures;
- Integration of crop-livestock-forest and agroforestry systems;
- Direct planting system;
- Biological nitrogen fixation;
- Planted forests;
- Treatment of animal waste;
- Adaptation to climate change.

This kind of strategies depends on the characteristics of each country (i.e., availability of soil and water, climate, value chains, etc.). However, we can note common opportunities for R&D and innovation, mainly those related with environmental chemistry to understand fate and dynamics. Moreover, goal 13 of the sustainable development is associated with the climate action because climate change is affecting every country on every continent [25].

Based on data and simulations carried out in different regions of the globe, the same IPCC [10] highlighted that global warming is unequivocal. Since 1950, unprecedented changes have been observed over decades or millennia: the atmosphere and the ocean have warmed, the layers of ice and snow have decreased and the level of the oceans has risen. Should GHGs emissions continue to rise at current rates, the planet's temperature could rise by 5.4 °C by 2100.

As a consequence, the sea level may rise up to 82 cm and impact most of the coastal regions. For instance, for countries as Brazil and others in South America, the main expected impacts are [3]:

- Extinction of habitats and species, mainly in the tropical region.
- Replacement of tropical forests by savannas and semi-arid vegetation by arid.
- Increase in regions in situations of water stress, that is, without enough water to meet the demands of the population.
- Increase of pests in agricultural crops.
- Increased incidence of diseases such as dengue and malaria, in addition to population migration.

Agriculture plays two roles in the climate change scenario. It is an activity that emits GHGs, which contributes to global warming, and it is an activity highly sensitive to climate change. This raises the need for low carbon agriculture and the development of technologies to mitigate the negative effects of the climate on crops and livestock and those negative effects from agroindustrial activities.

Water management

Nowadays, water suffers a high negative impact from agricultural activities due to its huge use for crops and livestock production and clean water and sanitation is the goal 6 of the United Nations' sustainable goals (United Nations 2020). World agriculture consumes approximately 70% of the fresh water withdrawn per year [24]. Moreover, its quality and potability is compromised by the presence of pesticides and emerging pollutants from livestock activities (Permission 2018).

The optimization of water uses is paramount for a sustainable agriculture, for that we can apply reuse and treatment strategies [27]—recycled water is of special interest for agricultural purposes, because decreases wastewater discharges and reduces and prevents pollution—associated to technologies for a reduction in the consumption, as the calculation of net irrigation based on effective crop water requirement [14].

Finally, there are some general strategies for water management in agriculture:

- Monitoring and control the presence of metals, organic chemicals and microorganisms in superficial water and groundwater;
- Establishment of security plans for the case of polluting events;
- Treatment plants for residual water, mainly those from pesticide application;
- Control of the pesticides, fertilizers, etc., uses in field;
- Reuse and recycling;

- Optimization.

As for CO_2 emission control, environmental chemistry is very relevant to maintain a water security because provides scientific and technological knowledge.

Figure 3.5 depicts a water management strategy applied to agriculture and agroindustry.

There are some considerations to take in mind regarding the content of the Fig. 3.5:

- Use of techniques to reduce water demand: in agriculture it can be related, for instance, to the use of drip system instead of large continuous water flows; in agroindustry it can be related to the decreasing in the water by the processing steps.
- Don't pollute! it is mandatory to improve or to maintain the water quality and it is closely related to the smart management of raw materials, reactants, products and residues/wastes.

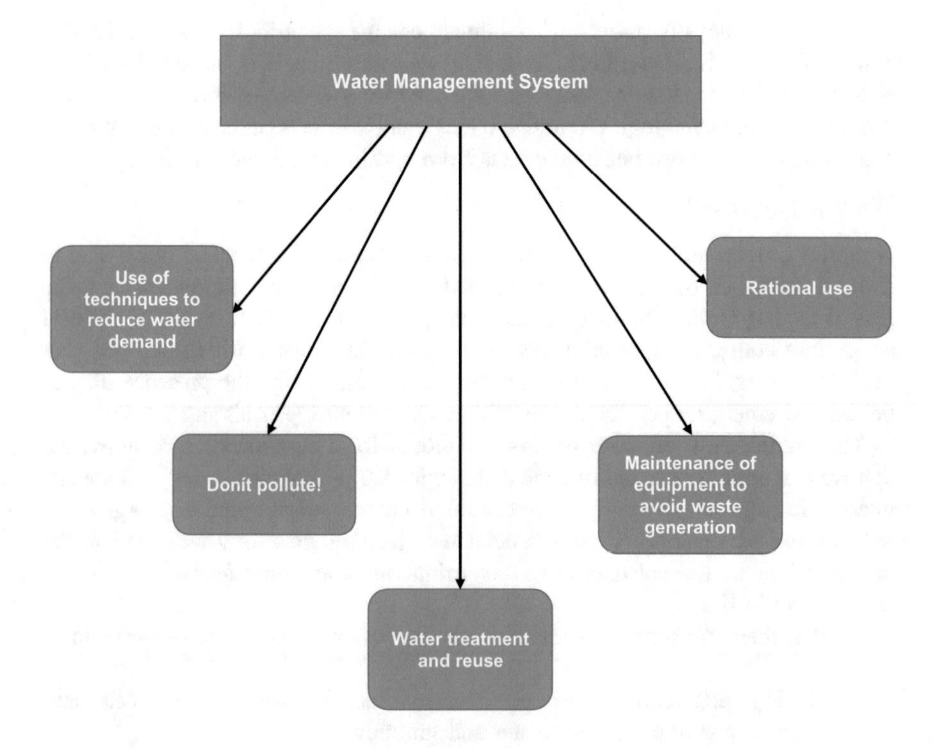

Fig. 3.5 Proposal of water management system. Author

- Water treatment and reuse: it is closely related to the application of water and wastewater treatment technologies—e.g., aerobic wastewater treatment, anaerobic digestion, clarification water, continuous electrodeionisation, deionization, filtration, incineration, industrial distillation, odour control, reverse osmosis, separation, sludge pump, thermal hydrolysis, ultrafiltration vacuum evaporation—and the cyclic use of the water, mainly by the agroindustry.
- Maintenance of equipment: old and damaged equipment generates more waste than new ones with up-to-date maintenance.
- Rational use: it comprises the conscious use of water as a finite and extremely valuable resource for the whole of society.

3.4 Conclusions

Environmental chemistry, as a branch of Chemistry and Environmental Sciences, can contribute with the support of knowledge regarding to the biogeochemical cycles of the chemical elements in the nature, mainly carbon, as well as the genesis and fate of organic matter, specially that obtained from agroindustrial biomass.

However, biomass production and processing generate environmental issues to be observed and understanding, related to greenhouse gas emissions and water pollution. Additionally, a reduction on CO_2 emission and the water management will guarantee sustainable strategies based on scientific statements.

References

1. Anderson TR, Hawkins E, Jones PD (2016) CO_2, the greenhouse effect and global warming: from the pioneering work of Arrhenius and Callendar to today's Earth system models. Endeavour 40:178–187
2. Brazilian Agricultural Research Corporation (2020a) Low-carbon emission agriculture. https://www.embrapa.br/tema-agricultura-de-baixo-carbono/sobre-o-tema. Accessed April 2020
3. Brazilian Agricultural Research Corporation (2020b) Mudança do clima (Climate change). https://www.embrapa.br/en/visao/mudanca-do-clima. Accessed June 2020
4. Burdon J (2001) Are the traditional concepts of the structures of humic substances realistic? Soil Sci 166:752–769
5. Clapp CE, Hayes MHB, Senesi N, Bloom PR, Jardine PM (eds) (2001) Humic substances and chemical contaminants. Soil Society of America, Madison
6. Diallo MS, Simpson A, Gassman P, Faulon JL, Johnson-Jr JH, Goddard WA, Hatcher PG (2003) 3-D structural modeling of humic acids through experimental characterization, computer assisted structure elucidation and atomistic simulations. 1. Chelsea soil humic acid. Environ Sci Technol 37:1783–1793

7. Food and Agriculture Organization of the United Nations (2014) Greenhouse gas emission from agriculture, forestry and other land use. http://www.fao.org/resources/infographics/infographics-details/en/c/218650/. Accessed April 2020
8. Food and Agriculture Organization of the United Nations (2020a) Wastewater treatment and reuse in agriculture. http://www.fao.org/land-water/water/water-management/wastewater/en/. Accessed April 2020
9. Food and Agriculture Organization of the United Nations (2020b) FAO soils portal. http://www.fao.org/soils-portal/soil-management/soil-carbon-sequestration/pt/. Accessed April 2020
10. Intergovernmental Panel on Climate Change (2014) Climate change 2014: mitigation of climate change. Contribution of working group III to the fifth assessment report of the Intergovernmental Panel on Climate Change. Cambridge University Press, Cambridge
11. International Union of Pure and Applied Chemistry (2020) IUPAC Gold book. https://goldbook.iupac.org/terms. Accessed April 2020
12. Jenkinson EJ, Adamns DE, Wild A (1991) Model estimates of CO_2 emissions from soil in response to global warming. Nature 351:304–306
13. Martín-Neto L, Tragghetta DG, Vaz CMP Jr, Crestana S, Sposito G (2001) On the interaction mechanisms of atrazine and hydroxyatrazine with humic substances. J Environ Quality 30:520–525
14. Organization for Economic Co-operation and Development (2009) The bioeconomy to 2030: designing a policy agenda. http://www.oecd.org/futures/long-termtechnologicalsocietalchallenges/thebioeconomyto2030designingapolicyagenda.htm. Accessed April 2020
15. Piccolo A (2001) The supramolecular structure of humic substances. Soil Sci 166:810–832
16. Piccolo A, Spaccini R, Savy D, Drosos M, Cozzolino V (2019) The soil humeome: chemical structure, functions and technological perspectives. In: Vaz S Jr (ed) Sustainable agrochemistry—a compendium of technologies. Springer Nature, Cham
17. Ruane AC, Phillips MM, Rosenzweig C (2018) Climate shifts within major agricultural seasons for +1.5 and +2.0 °C worlds: HAPPI projections and AgMIP modeling scenarios. Agric Forest Meteorol 259:329–344
18. Snoeyink VL, Jenkins D (1996) Química del água (Water chemistry). Limusa, México City
19. Stevenson FJ (1994) Humus chemistry: genesis, composition, reaction, 2nd edn. Willey, New York
20. Sutton R, Sposito G (2005) Molecular structure in soil humic substances: the new view. Environ Sci Technol 39:9009–9015
21. Swift RS (1999) Macromolecular properties of soil humic substances: fact, fiction, and opinion. Soil Sci 164:760–802
22. Tatzber M, Stemme M, Spiegel H, Katzlberger C, Hanernhauer G, Gerzabek MH (2008) Impact of different tillage practices on molecular characteristics of humic acids in a long-term field experiment—an application of three different spectroscopic methods. Sci Total Environ 406:256–268
23. Uhde E, Salthammer T (2007) Impact of reaction products from building materials and furnishings on indoor air quality—a review of recent advances in indoor chemistry. Atmos Environ 41:3111–3128
24. United Nations Educational, Scientific and Cultural Organization (2020) The 2020 world water development report. http://www.unesco.org/new/en/natural-sciences/environment/water/wwap/. Accessed April 2020
25. United Nations (2020) Sustainable development goals. https://www.un.org/sustainabledevelopment/sustainable-development-goals/. Accessed April 2020
26. United States Environmental Protection Agency (2020a) Greenhouse gas emissions. https://www.epa.gov/ghgemissions/global-greenhouse-gas-emissions-data. Accessed April 2020

27. United States Environmental Protection Agency (2020b) Water reuse and recycling. https://www3.epa.gov/region9/water/recycling/. Accessed April 2020
28. World Health Organization (2006) WHO guidelines for the safe use of wastewater, excreta and greywater—v. 2. Wastewater use in agriculture. World Health Organization, Geneva
29. World Health Organization (2017) Guidelines for drinking-water quality, 4th edn. World Health Organization, Geneva

An Outlook of Monitoring and Treatment Technologies

4

Abstract

Nowadays, we can observe a concern from society with the negative impacts to environment and public health from agriculture and agroindustry. It is becoming paramount the development of a more environmentally and healthy friendly agribusiness. To achieve it are necessary technologies of monitoring—as real-time monitoring and probes—and technologies of treatment—as chemical, biochemical, thermochemical and physical processes for the generated residues. This Chapter deals with the introduction of more relevant technologies of monitoring and treatment to be applied to agroindustrial residues, and the chemical composition of some of them. Furthermore, the proposal of a renewable carbon based-technologies for agroindustrial purposes is presented allied to adding value, biorefinery and bioeconomy concepts.

4.1 Polluting Aspects of Biomass from Agricultural Crop Practices and Agrofood Processing Industry

Governments, farmers and consumers shows increasing concerns with the negative impacts on the environment and health caused by the modern agriculture, as the large amount of inputs applied to produce different crops in different regions around world. Moreover, agrochemicals - those inputs used in the tillage systems—have a direct correlation with damages from agriculture, with pesticides being the main representative class with toxicological implications.

According Vaz Jr. [32], the main negative impacts on the environment from the agriculture are:

- Water, soil and air pollution due to pesticide applications;
- Extinction of water bodies, due to high water demand;

S. Vaz Jr., *Treatment of Agroindustrial Biomass Residues*,
https://doi.org/10.1007/978-3-030-58850-2_4

- Erosion and soil degradation due to inadequate management during tillage;
- Change in biota due to factors already listed;
- Changes in the quality of environmental resources, also due to factors already listed;
- Ecological risks for insects, plants and animals associated with the change of environment;
- Climate change due to the deforestation and biomass combustion.

Additionally, Vaz Jr. (2019a) also pointed as impacts on health:

- Poisoning due to pesticide use and food contaminated consumption;
- Occupational risks to farmers due to the exposition to pesticides;
- Human infections—or emerging infectious diseases, that do not respond to treatment due to the use of antimicrobials in agriculture [11].

From these negative impacts, it is becoming paramount the development of a more environmentally and healthy friendly agriculture and agroindustry.

Regarding to the plant biomass for agroindustrial purposes, Cantor and Rizy [7] proposed environmental risks associated to the biomass for energy, which can be extended to general agricultural residues:

- Deforestation
- Wood combustion
- Fertilizers, pesticides, herbicides
- Competition with food production
- Irrigation
- Ecological diversity
- Soil erosion
- Nutrient depletion
- Sedimentation
- Dust emissions

These risks have a closer relation with those main negative impacts previous listed above.

Then, the fact that the biomass residue originates from vegetables does not mean that it does not have direct or indirect risks associated with it. For instance, in the case of pulp and paper industry there are sawdust, chips, bark, and sludge as residues that can generate environmental negative impacts. Sadh et al. [25] observed that agricultural-based industries produce vast amount of residues every year, and the majority of the agro-industrial wastes are untreated and underutilized, being disposed either by burning, dumping or unplanned landfilling. That means, generating environmental liabilities for the society.

As established in the Chap. 2, impacts can be positive or negative according their direct or indirect effects on the environment, economy and society. Considering that a historic of incidents at global level, environmental impacts are expected to be negative when agroindustrial activities and systems are considered; nevertheless, it could be positive if modern technologies and good agricultural practices are used. A more detailed evaluation of sustainability in agriculture can be seen in Quintero-Angel and González-Acevedo [22] and Vaz Jr. [36].

Regarding to agrofood processing industry, there are huge variety and amount of residues, as wastewater effluents produced in a cheese factory; orange juice agrofood residues; effluent from olive oil, biomass waste from wine production, etc. Then, each processed biomass for food and feed has its own residues, as—but not limited to:

- Effluents (liquid and gaseous)
- Organic acids, phenols and polyphenols
- Sugars
- Oils
- Waxes and fats
- Cellulose, lignin and hemicellulose
- Proteins.

Furthermore, agrofood residues bring with them those residues historic from the agricultural or livestock production systems. Then, there is a sum of impacts to be considered in a life cycle assessment.

Parts of item 4.1 are reproduced with permission from Springer Nature.

4.2 Description of Technologies for Monitoring

Analytical technologies depict here were chosen according their large use or potential to use in the biomass production and processing.

It is worthy to observe that monitoring technologies should be associated to treatment technologies in order to obtain best results.

4.2.1 Real-Time Monitoring

This set of technologies allow the observation in real-time a certain production step (e.g., harvest) and a certain processing step (e.g., fermentation).

Saar et al. [24] applied stimulated raman scattering microscopy for a real time monitoring of biomass processing. The technique allowed the degradation of lignin in the cell wall to be monitored with high sensitivity, sub-micrometer spatial resolution, and high temporal resolution.

4.2.2 Spectroscopic Techniques

It is a large set of analytical technologies with some advantages in their use, as ease of handling.

Raman, a spectroscopic technique, was previously cited to real-time monitoring. Additionally, we can cite ^{1}H and $^{1}3C$ nuclear magnetic resonance to determine polyphenols in bark biomass residues; this technique was associated to liquid chromatography—HPLC-UV and HPLC-MS [27].

4.2.3 Probes and Sensors

Probes and sensors are specific devices ease to hand under field conditions, which electrochemical and spectroscopy are the main used analytical techniques or principles.

Camerani et al. [6] applied combined synchrotron radiation induced micro X-ray fluorescence (μ-SRXRF) and tomography (μ-SRXRFT) techniques to obtain information about Cd distribution inside single municipal solid waste and biomass fly ash particles, because it is fundamental since it affects the residue leachability. Once again, we have two spectroscopic techniques.

4.2.4 Process Analytical Chemistry

This concept encloses the other three sets of analytical technologies previously commented.

The need for quality control of processes for converting biomass into products, especially those of greater value, has boosted the advancement of process analytical chemistry (PAC)—often also known as PAT (process analytical technology). As an area of research and application, the use of robust techniques and methods is preferred, as well as in real time, with the analyses being carried out directly in the reactor or bioreactor, instead of analyses carried out in the laboratory. The main advantage of this type of analytical approach in relation to the traditional one, where manual sampling is carried out followed by sample transport and subsequent analysis in the laboratory, is that analyzes performed "in situ" provide greater speed for taking corrective actions and consequent adjustment of the production process.

However, some aspects must be carefully considered in the methodological planning of measures in dynamic systems [28] such as, among others:

- Selection of process variables, such as temperature, pressure and pH, or definition of the process quality parameters to be measured;
- Establishing a quantitative relationship between controllable properties, paying attention to the fact that there is not always a direct relationship between them.

4.2.5 Chemical Evaluation of Organic Residues

A classification of organic residues, as agricultural biomass, can be seen as a useful tool to control and to treat these residues by means a modern and practical approach.

A well-established chemical characterization of organic residues can use as reference the norm ISO/DIS 17300 (2020) for biogenic carbon–carbon derived from biomass, adapted for each circumstance and necessity.

Figure 4.1 depicts a simplified flowchart dedicated to the chemical evaluation in order to address the best uses of the organic residue.

A good and useful reference of analytical methods for biomass residues can be accessed free of charge on the National Renewable Energy Laboratory (NREL) website [20]. For instance:

- Preparation of samples for compositional analysis: describes a reproducible way to convert a variety of biomass samples into a uniform material suitable for compositional analysis.
- Determination of total solids in biomass and total dissolved solids in liquid process samples: intended to determine the amount of total solids remaining after 105 °C drying of a biomass sample.
- Determination of ash in biomass: applicable to hard and soft woods, herbaceous materials, agricultural residues, wastepaper, and solid fraction process samples.
- Determination of protein content in biomass: suitable for biomass feedstocks, process solids, and process liquids.
- Determination of extractives in biomass: two-step extraction process to remove water soluble and ethanol soluble material.
- Determination of structural carbohydrates and lignin in biomass: appropriate for extractives free biomass, as well as process solids containing no extractives.

Parts of the item 4.3 are reproduced with permission from National Renewable Energy Laboratory.

4.2.6 Chemical Composition of Some Agroindustrial Wastes and Residues

The determination of the chemical composition of biomass residues is paramount and the first step in the treatment processing. Generally, we have a solid residue (e.g., lignocellulosic biomass). However, we can also have a liquid or semi-solid effluent (e.g., pig slurry).

The chemical composition of some relevant agroindustrial wastes and residues are descripted here in order to support the Chap. 6, dedicated to strategies of treatment. For a better understanding of analytical methods for this kind of matrices, a book as "*Analytical Techniques and Methods for Biomass*" [33] can be used as reference.

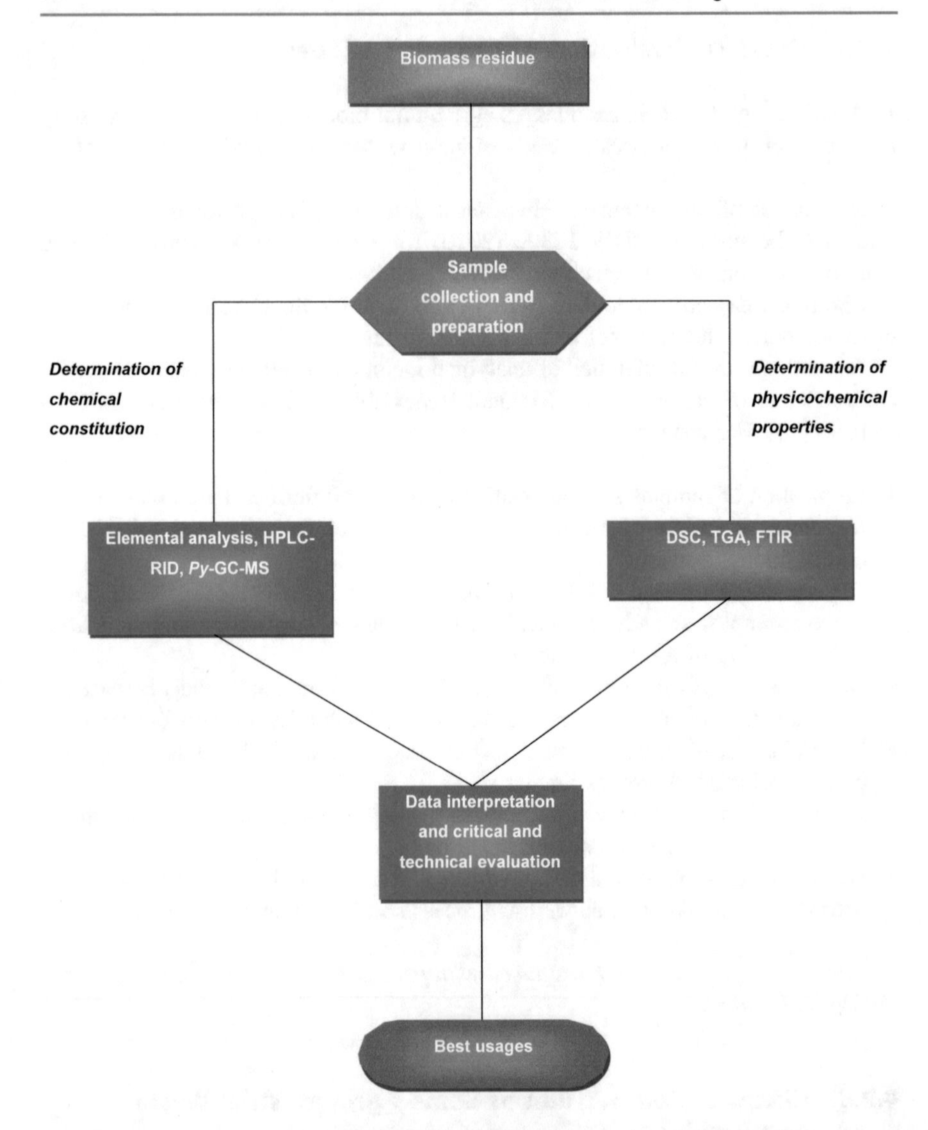

Fig. 4.1 A flowchart for the chemical evaluation of organic residues from agricultural sources. *HPLC-RID* high performance liquid chromatography with a refractive index detector; *Py-GC-MS* gas chromatography with a mass spectrometry detector and a pyrolysis probe; *DSC* differential scanning calorimetry; *TGA* thermal gravimetric analysis; *FTIR* Fourier transformed infrared spectroscopy. Author

Additionaly, analytical techniques as biochemical oxygen demand (BOD), chemical oxygen demand (COD) and total organic carbon (TOC) are very useful for liquid effluents. The first analyses the oxygen consumed by microorganisms presented in water/effluent sample; the second the oxygen consumed by organic matter presented also in water/effluent sample; the third analyses the *total amount of carbon in organic compounds in water/effluents and applied to determine how suitable an aqueous solution is for their processes—e.g., a treatment involving carbon conversion from organic to inorganic, and* vice versa. For a better understanding of analytical methods dedicated to these analytical techniques the best reference book is "Standard Methods for the *Examination* of *Water and Wastewater" [1].*

Black liquor from pulp & paper industry

According Xu et al. [38], the chemical composition of a black liquor from pulp and papel industry is:

- Total solids (g L^{-1}): 51.7
- Volatile solid (g L^{-1}): 23.7
- pH: 14.09
- Alkalinity (g NaOH/L): 12
- Carbohydrates (g L^{-1}): 4.37
- Total reducing sugars (g L^{-1}): 0.44
- Total lignin (g L^{-1}): 9.45
- Acid insoluble lignin (g L^{-1}): 3.57
- Electrical conductivity (ms cm^{-1}): 61.3
- Na (g L^{-1}): 15.83
- K (g L^{-1}): 1.49
- Ca (g L^{-1}): 1.25
- Mg (g L^{-1}): 0.93.

Orange peel

According Lachos-Perez et al. [19], the chemical composition of orange peel is:

- Moisture (% wt./wt.): 8.71 ± 0.03
- Ash (% wt./wt.): 3.99 ± 0.04
- Protein (% wt./wt.): 6.85 ± 0.00
- Ethanol extractibles (% wt./wt.): 17.15 ± 1.73
- Cyclohexane extractibles (% wt./wt.): 3.79 ± 1.04
- Lignin, soluble (% wt./wt.): 7.71 ± 3.58
- Lignin, insoluble (% wt./wt.): 9.09 ± 1.53
- Pectin (% wt./wt.): 19.62 ± 3.24
- Glucan (% wt./wt.): 34.22 ± 4.68
- Xylan (% wt./wt.): 2.15 ± 0.29
- Arabinan (% wt./wt.): 4.72 ± 0.88
- Carbon (% wt./wt.): 42
- Hydrogen (% wt./wt.): 6

- Nitrogen (% wt./wt.): 1
- Oxygen (% wt./wt.): 51
- Sulfur (% wt./wt.): 0.

Pig slurry

According Hunce et al. [13], the chemical composition of a pig slurry in dry basis is:

- Moisture (% wt./wt.): 62.7 ± 7.04
- pH: 7.49 ± 0.25
- Electrical conductivity (dS m^{-1}): 6.10 ± 0.82
- Organic matter (% wt./wt.): 32.0 ± 5.14
- Water soluble carbon (g kg^{-1}): 8.0 ± 0.27
- Total organic carbon (g kg^{-1}): 381 ± 19.2
- Total nitrogen (g kg^{-1}): 22.1 ± 1.90
- C/N: 17.3 ± 1.79
- Soluble carbohydrates (g kg^{-1}): 2.7 ± 0.0.

Sugarcane bagasse

According Vassilev et al. [30], the major composition of sugarcane bagasse is:

- Cellulose (%wt./wt.): 42.7
- Hemicellolose (%wt./wt.): 33.1
- Lignin (% wt./wt.): 24.2.

Waste of beer fermentation

According Khan et al. [18], the chemical composition of a waste of beer fermentation broth is:

- Carbon (g L^{-1}): 68.114
- Nitrogen (g L^{-1}): 17.178
- Hydrogen (g L^{-1}): 116.141
- Total carbohydrate (g L^{-1}): 29.442
- Total protein (g L^{-1}): 0.671
- Ethanol (% v/v): 4.58.

Wood residues processed in paper & pulp industry

According Ribeiro et al. [23], the chemical composition of *E. urophylla* x *E. grandis* eucalyptus hybrid wood is:

- Basic density (kg m^{-3}): 452 ± 1.40
- Ash (% wt./wt.): 0.26 ± 0.02

- Extractives in acetone (% wt./wt.) 1.30 ± 0.03
- Acid-insoluble lignin (% wt./wt.): 24.0 ± 0.00
- Acid-soluble lignin (% wt./wt.): 4.10 ± 0.01
- Total lignin (% wt./wt.): 28.1 ± 0.01
- Lignin S/G ratio: 2.90 ± 0.07
- Acetyl groups (% wt./wt.): 2.10 ± 0.14
- Total uronic acids (% wt./wt.): 2.80 ± 0.07
- Glucans (% wt./wt.): 52.1 ± 0.13
- Xylans (% wt./wt.): 11.6 ± 0.15
- Galactans (% wt./wt.): 1.30 ± 0.06
- Mannans (% wt./wt.): 0.80 ± 0.01
- Arabinans (% wt./wt.): 0.20 ± 0.00
- Total sugar (% wt./wt.): 66.0 ± 0.10
- Cellulose (% wt./wt.): 51.3 ± 0.10
- Hemicellulose (% wt./wt.): 19.6 ± 0.10.

4.3 Description of Technologies for Treatment

The use of agroindustrial wastes as raw materials can help to reduce the production cost in the productive chain and also reduce the pollution load from the environment—high organic matter content can promote eutrophication of waters; moreover, it is as an opportunity to aggregate value. However, to achieve these statements it is necessary the development and the application of a set of technologies according the residue chemical and physical characteristics. Moreover, and as cited before, these technologies should be associated those monitoring technologies for best results.

As mentioned in the item 5.2, those technologies presented here will be considered in details in the Chap. 5, because the subject of this item is introduce technologies with a high potential of application, considering four categories of processes: chemical, biochemical, physical and thermochemical.

An advantage of these processes is they could be seen as green processing technologies, because they work with the use of renewable feedstock and catalysts, the reduction in the energy demand, among other industrial friendly factors, in a closer relation with the green chemistry principles [2]. Indeed, the use of biomass and its residues as a renewable raw material can contribute in the practical application of the green chemistry principles, and to promote sustainable production chains [35].

4.3.1 Chemical Processes

These processes are commonly based on the use of catalysts—heterogeneous (metals, zeolites, immobilized enzymes) or homogeneous (acids, alkalis and dissolved salts)—to convert biomass residues into products as chemicals and materials. It is frequently associated with the utilization of lignocellulosic wastes.

As an example of this category, Shikinaka et al. [26] reported a simple yet effective method for processing lignocellulosic biomass by combined wet-type ultrafine bead milling and enzymatic saccharification at pH 5.0 and 50 °C. This generated nanoscale particles that allow close to 70% saccharification of cellulose and recovery of a glassy, flame retardant, and transparent non-deteriorated lignin-rich film.

4.3.2 Biochemical Processes

Probably, this is the main category of processes applied to the treatment of biomass residues due to the fact that a high content of organic matter of the last. Consequently, it has a high content of carbon, oxygen and nitrogen for metabolic processes of several microorganisms, as bacteria, yeast and fungi. As well observed by Cecconet et al. [8], agrofood wastes have a high biodegradability derived from their chemical constitution.

We can highlight the fermentation processes as the main bio-strategy to convert residues into end-products, as biogas, biofertilizers, organic acids, alcohols, among others. For instance, Fang et al. [10] applied the solid-state fermentation (SSF), using corn stalk and soybean meal as solid substrates to produce poly-γ-glutamic acid (γ-PGA) by *Bacillus amyloliquefaciens* JX-6. Furthermore, Haosagul et al. [12] produced biomethane (CH_4) from co-digestion process of waste glycerol and banana wastes; the microbial sludge was collected from the up-flow anaerobic sludge blanket (UASB) process treating cassava processing wastewater.

4.3.3 Thermochemical Processes

Thermochemical processes are a large category that involves torrefaction, carbonization, pyrolysis and gasification, with the biomass transformation into products occurring by means a thermal energy—sum of kinetic and potential energies—application. They have a large application to lignocellusic residues.

For instance, Kanwal et al. [17] applied the torrefaction process to sugarcane bagasse, obtaining a product comparable with charcoal for energetic purposes from crops and forests.

4.3.4 Physical Processes

Generally, physical processes are used in previous step for residues treatment, as grinding, drying, particle size reduction and compaction. They performs in association with those other process categories.

Chen et al. [9] used the granulation of agricultural bio-waste and its performance as a solid biofuel during the combustion process in comparison with two standard wood fuels. It was observed that the granular bio-waste fuels had excellent flowability during the combustion process.

4.4 Renewable Carbon and Its Relavance

Despict a concern with the CO_2 generation, amount of emission and its effects on the environment, C-derivatives, e.g., petrochemicals, are a raw material of several ending-products consumed by the modern society. In this way, is desirible the exchange of a C-fossil, non-renewable, by a C-biomass, renewable. Moreover, the own CO_2 generated by an industrial plant can be used as feedstock to produce inorganic carbonates and polymers, among other products whose will contribute for the sustainability of the respective chain of production.

Renewable carbon from biomass and its residues can be used for (University of Tenneessee 2020):

- Materials
- Chemicals
- Fuels
- Power
- Manufacturing

Then, renewable carbon from biomass source can auxiliary in order to achieve a more sustainable processing industry.

4.5 Adding Value to Biomass Residues

The *value-added concept* is an economic idea that comprises an increase in the value of certain raw material by means its processing. In other words, the more processed a raw material is, is expected that the greater the value (or price) it and its final products will have. Figures 4.2 and 4.3 illustrate the chain of products from the industrial processing of two crops; in this case, soybean and sugarcane, respectively.

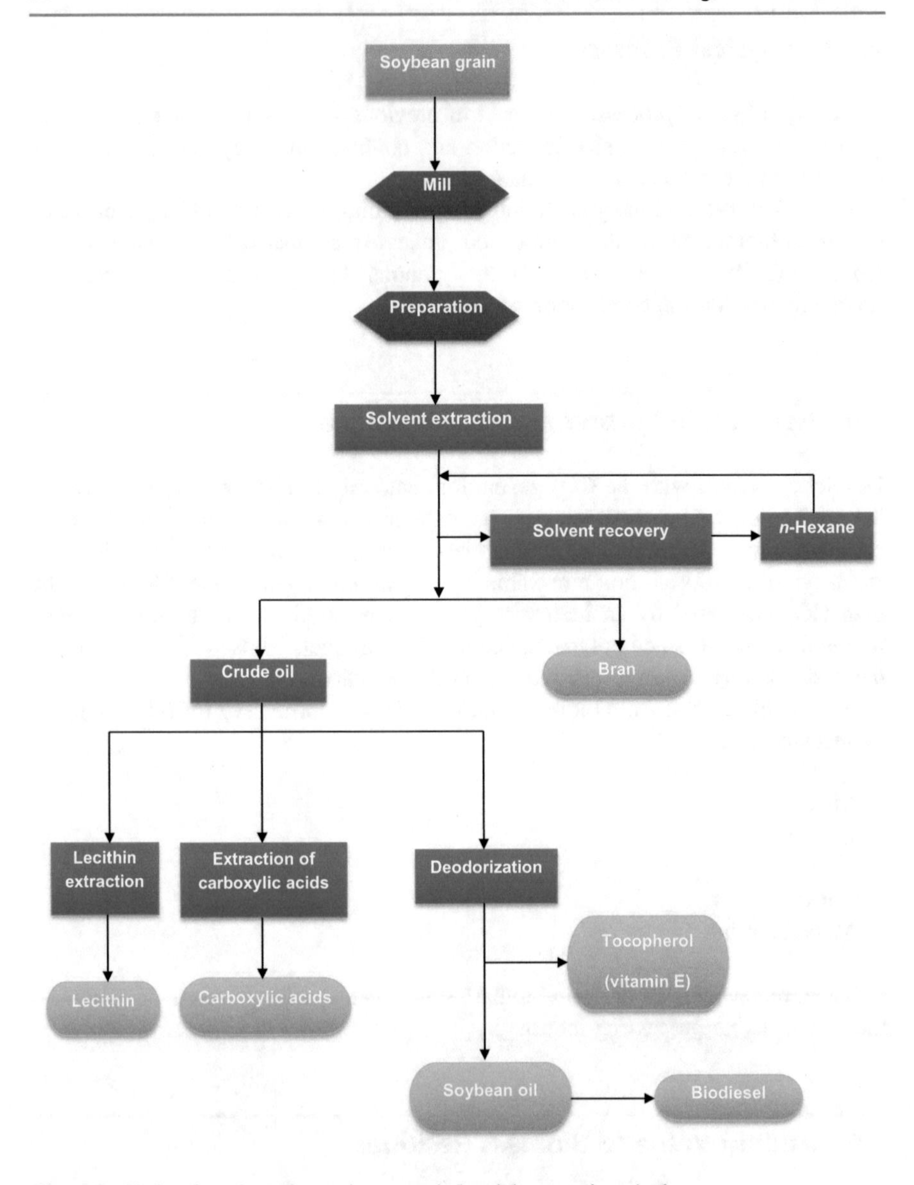

Fig. 4.2 Chain of products for soybean agroindustrial processing. Author

Taking as example the Fig. 4.2, we have a *biorefinery* of soybean—an industrial facility according the oil refinery model for a maximum exploitation of the raw material: the final products can be used as a source of protein for animal feed (bran), in the pharmaceutical and cosmetic industry (tocopherol), food (refined oil and lecithin), surfactants (carboxylic acids) and biofuel (biodiesel). Furthermore, there is the residue from pressing, cake, and the co-product from biodiesel synthesis,

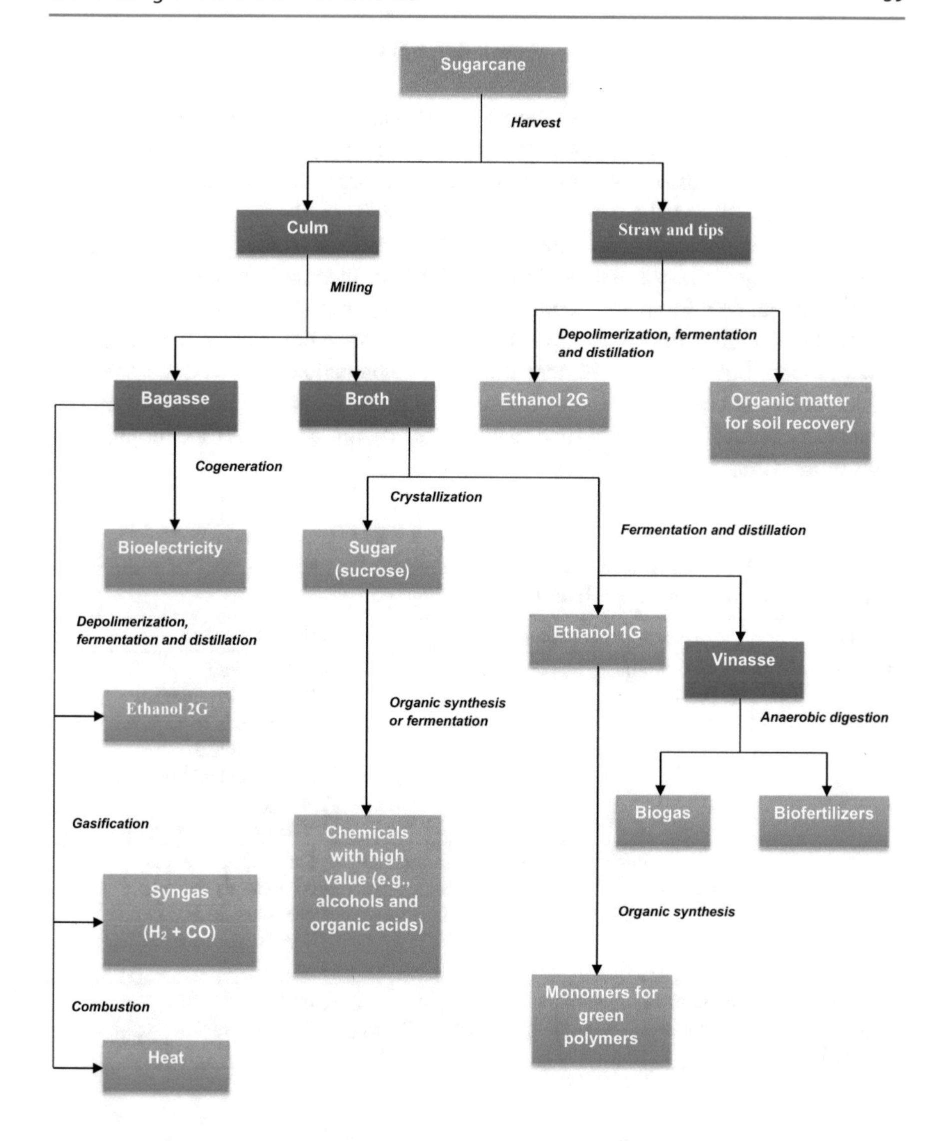

Fig. 4.3 Chain of products for sugarcane agroindustrial processing. *Source* adapted from Vaz Jr. [34]. Reproduced with permission from Springer Nature

glycerin. Every day new products and applications are being sought for biomass, so that the proposal for a "green" alternative to petrochemicals can be effectively viable taking into account technical and economic limitations inherent to this subject. An example of this is the glycerin obtained as a co- or by-product of biodiesel production; in addition to its wide use in the cosmetics industry, its use is sought to obtain monomers, such as acrylic acid, which can be used in the production of "green" polymers. Figure 4.3 follows the same approach, i.e., the whole industrial use of the biomass.

Agroindustrial wastes and residues represent one of the main potential sources of supply of raw material for renewable chemistry—a new paradigm in chemistry that uses renewable feedstock despite non-renewable ones, due to its large quantity produced and its well-established chains, which facilitates their processing. Tables 4.1 and 4.2 show the potential of the residues of the main bioenergetic chains of today—ethanol and biodiesel. It is worth mentioning that when looking for more noble uses for these residues, the same logic of the oil industry is followed, that is, the diversification of the potential of the raw material leads to a branching of new products from their constituent fractions. Lignocellulosic residues reflect this model well: the cellulose, hemicellulose and lignin fractions can be used as precursors to chemical compounds and materials according processing technologies, as seen in the Chap. 5.

Table 4.1 Residues from ethanol of sugarcane with potential to add value. Adapted from: Vaz Jr. [31]. Reproduced with permission from De Gruyter

Residue	Major chemical constitution	Proposal of use
Bagasse	Lignin, cellulose, hemicellulose, inorganics and water	Animal feed Renewable chemical compounds to replace petrochemicals Second generation (2G) ethanol Various alternative materials
Straw	Lignin, cellulose, hemicellulose, inorganics and water	Renewable chemical compounds to replace petrochemicals Second generation (2G) etanol
Vinasse (aqueous effluent)	Solubilized organic matter, insoluble inorganic solids, soluble inorganic salts and water	Bioga Biofertilizer

Table 4.2 Residues from biodiesel of soybean with potential to add value. Adapted from: Vaz Jr. [31]. Reproduced with permission from De Gruyter

Residue	Major chemical constitution	Proposal of use
Lignocellulosic residues	Lignin, cellulose, hemicellulose, proteins, inorganics and water	Animal feed Polymeric materials
Cake	Lignin, cellulose, hemicellulose, various organic compounds (proteins, esters, etc.), olefins and water	Animal feed
Aqueous effluent	Solubilized organic matter, insoluble and soluble inorganic solids, olefins and water	Biogas Biopolymer, e.g., polyhydroxyalkanoates (PHAs)

To strengthen, Jin et al. [16] observed that plant-derived wastes are good feedstocks for obtaining valuable compounds, e.g.:

- Proteins
- Pectin
- Polyphenols
- Essential oils
- Organic acids.

Then, it is correct claim that biomass and its residues are a source of sustainable wealth for the modern society.

The economic scenario for the world market for chemical products involves values of around USD 100 billion per year, where about 3% of this amount concerns bioproducts, or derivatives of biomass, with an estimate of an increase in this total participation to 25% up to the year 2025 [37]. These values give an idea of the possibilities and risks involved. For the case of chemical specialties and fine chemistry, the current share of renewables in about 25%, for both segments, may reach 50%, while for polymers the current 10% may reach 20%, also in 2025 [4].

By becoming an alternative for the use of renewable raw materials in conventional chemistry, fine chemistry and specialties, residual biomass can contribute to the reduction of negative environmental impacts of the production of chemical compounds, strongly dependent on oil, especially in the case of polymers. It is important to consider that the concept of biorefinery proposes the full use of the potential of biomass, following the model of an oil refinery to obtain energy, inputs, materials and chemicals. The fundamentals of *green chemistry* [3] establishes, among other criteria, the minimization of waste generation, the use of catalysts, energy and atomic savings and the use of renewable raw materials in chemistry; the *bioeconomy* [21] proposes to change an economy based on non-renewable resources, such as oil, for renewable resources, such as biomass.

It is possible to understand the economic potential of renewable chemistry through a pyramid that represents values and quantities for products (Fig. 4.4). In it, fine chemistry and specialties are at the highest level of adding value to biomass, which is defined according to the model of a biorefinery.

In a simple way, and according the content of Fig. 4.4, we can understand a biorefinery as follows:

(a) The biomass residues are processed in a biorefinery facility—processes can be chemical, biochemical, thermochemical or physical or an association among them (seen in the Chap. 5);
(b) In a first stage of valorization, energy and biofuels can be obtained and used by the own biorefinery; the surplus production can be commercialized;
(c) In a second stage of valorization, food and feed can be obtained and commercialized.
(d) In a third stage of valorization, chemical inputs and materials can be obtained and commercialized.

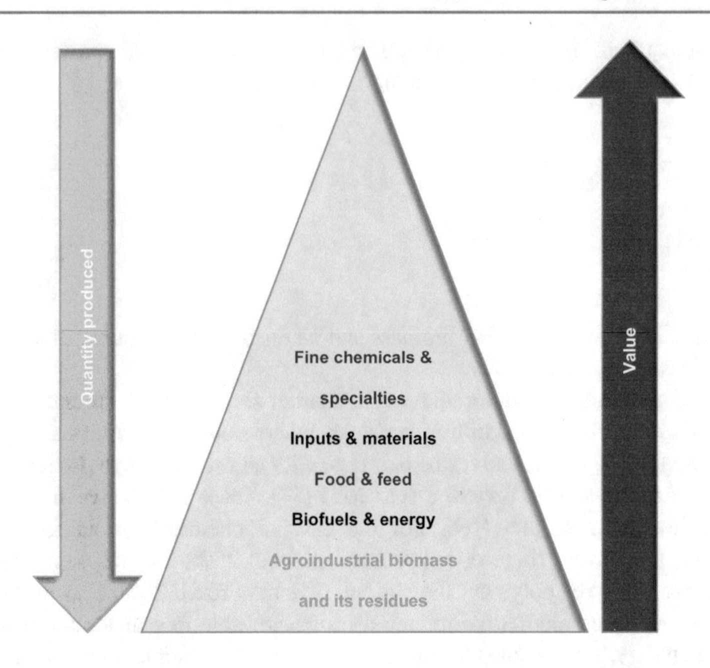

Fig. 4.4 Pyramid of values for agroindustrial biomass products and their residues according the biorefinery proposal. Author

(e) In a fourth and more profitable stage of valorization, fine chemicals and specialties can be obtained and commercialized.

Then, we can observe that the use of biomass residues from agroindustrial sources has a strong potential to create new value chains according the sustainability of products and processes—a set of sustainable practices. As an example of a biorefinery, we can refer the Borregaard's model, a wood-based biorefinery, developed to obtain sustainable products as specialty cellulose, lignin, vanillin and bioethanol [5]. Figure 4.5 shows this approach and its ending-products.

The International Energy Agency [14] performed a compilation of renewable (or bio-based) chemicals from biomass with a high potential to add value, comparing them against their petrochemical reference (Table 4.3).

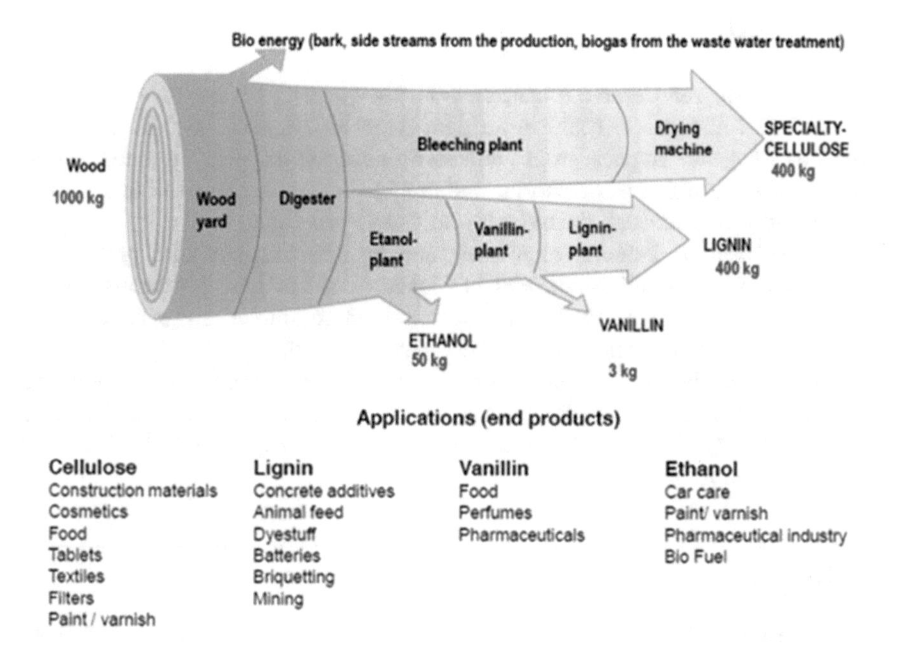

Fig. 4.5 Flowchart of products from the Borregaard's biorefinery approach. Courtesy of Borregaard

Table 4.3 Bio-based chemicals assessed for market penetration and reference materials. Source: adapted from International Energy Agency [14]. Reproduced with permission from International Energy Agency

Bio-based chemical	Reference petrochemical
Acetic acid	Acetic acid
Adipic acid	Adipic acid
n-Butanol	*n*-Butanol
Ethylene	Ethylene
Bio-MEG	MEG
Ethyl lactate	Ehtyl acetate
FDCA	Terephthalic acid
PHA	HDPE
PLA	PET and PS
Succinic acid	Maleic anhydide

MEG Monoethylene glycol; *FDCA* furan-2,5-dicarboxylic acid; *PHA* 3-[(21S,22S)-16-ethenyl-11-ethyl-4-hydroxy-3-methoxycarbonyl-12,17,21,26-tetramethyl-7,23,24,25-tetrazahexacyclo [18.2.1.15,8.110,13.115,18.02,6]hexacosa-1,3,5,8(26),9,11,13(25),14,16,18(24),19-undecaen-22-yl]propanoic acid; *HDPE* high density polyethylene; *PLA* 4-[3-[1-benzhydryl-5-chloro-2-[2-[[2-(trifluoromethyl)phenyl]methylsulfonylamino]ethyl]indol-3-yl]propyl]benzoic acid; *PET* polyethylene terephthalate; *PS* polystyrene

4.6 Conclusions

Nowadays, we can observe a concern from society with the negative impacts to environment and public health from agriculture and agroindustry. To minimize risks and negative impacts from biomass residues are necessary technologies of monitoring—as real-time monitoring and probes—and technologies of treatment—as chemical, biochemical, thermochemical and physical processes.

The association between monitoring and treatment technologies promotes a reduction in the pollution potential of biomass residues and the valorisation of them. Additionally, a renewable carbon from biomass source as raw material can auxiliary in order to achieve a more sustainable agroindustry.

The use of biomass residues from agroindustrial sources can promote the development of new value chains (e.g., chemicals, materials and energy) adding value to the agroindustry by means, for instance, the application of biorefinery concept allied to bioeconomy.

References

1. American Public Health Association, American Water Works Association, Water Environment Federation (2018) Standard Methods for the Examination of Water and Wastewater. Ed. 2018. https://www.standardmethods.org/. Accessed April 2020
2. American Chemical Society (2020) 12 principles of green chemistry. https://www.acs.org/content/acs/en/greenchemistry/principles/12-principles-of-green-chemistry.html. Accessed April 2020
3. Anastas PT, Warner JC (1998) Green chemistry: theory and practice. Oxford University Press, New York, NY
4. Biotechnology Industry Organization (2020). Biobased chemicals and products: a new driver for green jobs. http://www.bio.org/articles/biobased-chemicals-and-products-new-driver-green-jobs. Accessed June 2020
5. Borregaard (2020) The world's leading biorefinery. https://www.borregaard.com/Sustainability/Green-Room/The-world-s-leading-biorefinery. Accessed April 2020
6. Camerani MC, Golosio B, Somogyi A, Simionovici AS, Steenari B-M, Panas I (2004) X-ray fluorescence tomography of individual municipal solid waste and biomass fly ash particles. Analytical Chem 76:1586–1595
7. Cantor RA, Rizy CG (1991) Biomass energy: exploring the risks of commercialization in the United States of America. Bioresource Technol 35:1–13
8. Cecconet D, Molognoni D, Callegari A, Capodaglio AG (2018) Agro-food industry wastewater treatment with microbial fuel cells: Energetic recovery issues. Int J Hydrogen Energy 43:500–511
9. Chen H, Forbes EGA, Archer J, De Priall O, Allen M, Johnston C, Rooney D (2019) Production and characterization of granules from agricultural wastes and comparison of combustion and emission results with wood based fuels. Fuel 256:115897
10. Fang J, Liu Y, Huan CC, Xu L, Ji G, Yan Z (2020) Comparison of poly-γ-glutamic acid production between sterilized and non-sterilized solid-state fermentation using agricultural waste as substrate. J Cleaner Prod 255:120248
11. Grace D (2019) Infectious diseases and agriculture. Reference Module in Food Science. Encyclopedia Food Secur Sustain 3:439–447

12. Haosagul S, Boonyawanich S, Pitutpaisal N (2019) Biomethane production from co-fermentation of agricultural wastes. Int J Hydrogen Energy 44:5355–5364
13. Hunce SY, Clemente R, Bernal MP (2020) Selection of Mediterranean plants biomass for the composting of pig slurry solids based on the heat production during aerobic degradation. Waste Manage 104:1–8
14. International Energy Agency (2020) Bio-based chemicals—a 2020 update. https://www.ieabioenergy.com/wp-content/uploads/2020/02/Bio-based-chemicals-a-2020-update-final-200213.pdf. Accessed April 2020
15. International Organization for Standardization. Norm ISO/DIS 17300. Wood residue and post-consumer wood. https://www.iso.org/obp/ui#iso:std:iso:17300:-1:dis:ed-1:v1:en. Accessed April 2020
16. Jin Q, Yang L, Poe N, Huang H (2018) Integrated processing of plant-derived waste to produce value-added products based on the biorefinery concept.Trends Food Sci Technol 74:119–131
17. Kanwal S, Chaudhry N, Munir S, Sana H (2019) Effect of torrefaction conditions on the physicochemical characterization of agricultural waste (sugarcane bagasse). Waste Manage 88:280–290
18. Khan T, Hyon SH, Park JK (2007) Production of glucuronan oligosaccharides using the waste of beer fermentation broth as a basal medium. Enzyme Microbial Technol 42:89–92
19. Lachos-Perez D, Baseggio AM, Torres-Mayanga PC, Ávila PF, Tompett GA, Marostica M, Goldbeck R, Timko MT, Rostagno M, Martinez J, Forster-Carneiro T (2020) Sequential subcritical water process applied to orange peel for the recovery flavanones and sugars. J Supercritical Fluids 160:104789
20. National Renewable Energy Laboratory (2020) Laboratory analytical procedures. https://www.nrel.gov/bioenergy/laboratory-analytical-procedures.html. Accessed April 2020
21. Organization for Economic Co-operation and Development (2018) The bioeconomy to 2030: designing a policy agenda. http://www.oecd.org/futures/long-termtechnologicalsocietal challenges/thebioeconomyto2030designingapolicyagenda.htm. Accessed April 2020
22. Quintero-Angel M, González-Acevedo A (2018) Tendencies and challenges for the assessment of agricultural sustainability. Agric Ecosyst Environ 254:273–281
23. Ribeiro RA, Júnior SV, Jameel H, Chang H-M, Narron R, Jiang X, Colodette JL (2019) Chemical study of kraft lignin during alkaline delignification of *E. urophylla* x *E. grandis* hybrid in low and high residual effective alkali. ACS Sustain Chem Eng 7:10274–10282
24. Saar BG, Zeng Y, Freudiger CW, Liu Y-S, Himmel ME, Xie XS, Ding S-Y (2010) Label-free, real-time monitoring of biomass processing with stimulated raman scattering microscopy. Angewandte Chemie (International Ed. in English) 49:5476–5479
25. Sadh PK, Duhan S, Duhan JS (2018) Agro-industrial wastes and their utilization using solid state fermentation: a review. Bioresources Bioprocessing 5:1. https://doi.org/10.1186/s40643-017-0187-z
26. Shikinaka K, Otsuka Y, Navarro RR, Nakamura M, Shimokawa S, Nojiri M, Tanigawa R, Shigehara K (2016) Simple and practical process for lignocellulosic biomass utilization. Green Chem 18:5962–5966
27. Da Silva FMA, Hanna ACS, De Souza AA, Filho FAS, Canhoto OMF, Maganhães A, Benevides PJC, De Azevedo MBM, Siani AC, Pohlit AM, De Souza ADL, Koolen HHF (2019) Integrative analysis based on HPLC-DAD-MS/MS and NMR of *Bertholletia excelsa* bark biomass residues: determination of ellagic acid derivatives. J Brazilian Chem Soc 30:830–836
28. Van Staden JF (1999) Analytical aspects of chemical process control. Part 1. Fundamentals. Pure Appl Chem 71:2303–2308
29. University of Tennessee (2020) Center for renewable carbon. https://ag.tennessee.edu/crc/Pages/default.aspx. accessed April 2020
30. Vassilev SV, Boxter D, Andersen LK, Vassileva CG, Morgan TJ (2012) An overview of the organic and inorganic phase composition of biomass. Fuel 94:1–33

31. Vaz S Jr (2014) Perspectives for the Brazilian residual biomass in renewable chemistry. Pure Appl Chem 86:833–842
32. Vaz S Jr (2019) Introduction to sustainable agrochemistry. In: Vaz S Jr (ed) Sustainable agrochemistry—a compendium of technologies. Springer Nature, Cham
33. Vaz S Jr (ed) (2016) Analytical techniques and methods for biomass. Springer Nature, Cham
34. Vaz S Jr (2017) Sugarcane-biorefinery. In: Wagemann K, Tippkötter N (Eds) Biorefineries. Advances in biochemical engineering/biotechnology, vol 166. Springer, Cham
35. Vaz S Jr (ed) (2018) Biomass and green chemistry—building a renewable pathway. Springer Nature, Cham
36. Vaz S Jr (ed) (2019b) Sustainable agrochemistry—a compendium of technologies. Springer Nature, Cham
37. Vijayendran BJ (2010) Bio products from biorefineries—trends, challenges and opportunities. J Bus Chem 7:109–115
38. Xu H, Li Y, Hua D, Zhao Y, Mu H, Chen H, Chen G (2020) Enhancing the anaerobic digestion of corn stover by chemical pretreatment with the black liquor from the paper industry. Bioresource Technol 306:123090

Strategies of Treatment

5

Abstract

Agricultural biomass wastes/residues are predominantly crop stalks, leaves, roots, fruit peels and seed/nut shells that are normally discarded or burned. However, there are some challenges in trying to determine the extent of crop-produced biomass in relation to what is a 'loss' (from production, post harvesting and processing), or a 'waste' (retail or consumer loss). Then, the choice of a treatment strategy is not a simple task; in some cases, will be necessary an association of processes—chemical, biochemical, thermochemical and physical categories—to achieve the better result. This Chapter explore application strategies for these four processing categories to several agroindustrial wastes and residues. Moreover, aspects of environmental management and circular economy are discussed (Parts of this chapter were reproduced with permission from: Do Amaral AC, Kunz A, Steinmetz RLR, Contelli F, Scussiato LA, Justi KC (2014) Swine effluent treatment using anaerobic digestion at different loading rates. Engenharia Agrícola 34: 567–576. Copyright information: 2014, Scielo), (Parts of this chapter were reproduced with permission from: Garg A, Mishra IM, Chand S (2010) Effectiveness of coagulation and acid precipitation processes for the pre-treatment of diluted black liquor. Journal of Hazardous Materials 180: 158–164. Copyright information: 2010, Elsevier).

5.1 Available Technologies and Strategies of Application

According Tripathi et al. [26], agricultural biomass wastes/residues are predominantly crop stalks, leaves, roots, fruit peels and seed/nut shells that are normally discarded or burned but are in practice a potential valuable supply of feed-stock material. However, the Food and Agriculture Organization of the United Nations

S. Vaz Jr., *Treatment of Agroindustrial Biomass Residues*,
https://doi.org/10.1007/978-3-030-58850-2_5

[10] observed that there are some challenges in trying to determine the extent of crop-produced biomass in relation to what is a 'loss' (from production, post harvesting and processing), or a 'waste' (retail or consumer loss). Then, the choice of a treatment strategy is not a simple task; in some cases, will be necessary an association of processes to achieve the better result.

Figure 5.1 depict typical components of an economic exploitation of a certain biomass chain (e.g., grains), with the generation and harnessing of residues.

As presented in the thsi chapter, we can consider four categories of processes of treatment for these residues: chemical, biochemical, thermochemical and physical. Each category is described below. Moreover, these categories will be used in the biorefinery facilities—seen in the Chap. 4.

5.1.1 Chemical Processing

The chemical transformation or conversion processes are based on chemical reactions and, in most cases, a component of the biomass is extracted and purified, and later used as a starting reagent in a synthetic route, which often uses catalysts to increase the yield of the product of interest and to reduce reaction times. Thus, it is possible to note that several aspects of green chemistry [1], such as the use of catalysts and the reduction of waste generation, can be applied here, and the first aspect can become an extremely strategic item for this category of processes.

In the case of the use of cellulose and hemicellulose from the lignocellulosic residue, these polymers and their constituent sugars must be obtained first, with emphasis on glucose (hexose) and xylose (pentose), respectively, for the subsequent production of molecules of industrial interest [15]. For the case of lignin, an example of sought technology is the breakdown of its molecular structure, in order to mainly release phenolic compounds, which can be tested, for example, as monomers in different preparation routes—the formation of non-polar compounds can be obtained after its structural cracking, which will depend on the types of reaction and catalysts. Moreover, lignin can be used without any structural modification as ending-products [25].

Obtaining block-building compounds, which originate a large number of other compounds of economic interest, and synthesis intermediates, which can be used in fine chemistry, is the usual approach for research & development & innovation (R&D&I) projects [5, 27, 28]. Building-block compounds, such as furfural and xylitol (from the xylose that constitutes hemicellulose) and hydroxymethylfurfural (from glucose), among others, can add great value to carbohydrates from those residues [5, 15], with the same being possible to extend to resideual lignin from pulp and paper industry and glycerin from biodiesel production.

Figure 5.2 illustrates in a simplified way the application of chemical processes in the development of technologies for the use of co-products and waste. Initially, the biomass residue must undergo a complete chemical characterization, which aims to determine its chemical constitution, in addition to some physicochemical properties that are of interest. Then, there is the pre-treatment step of the residue, when it is

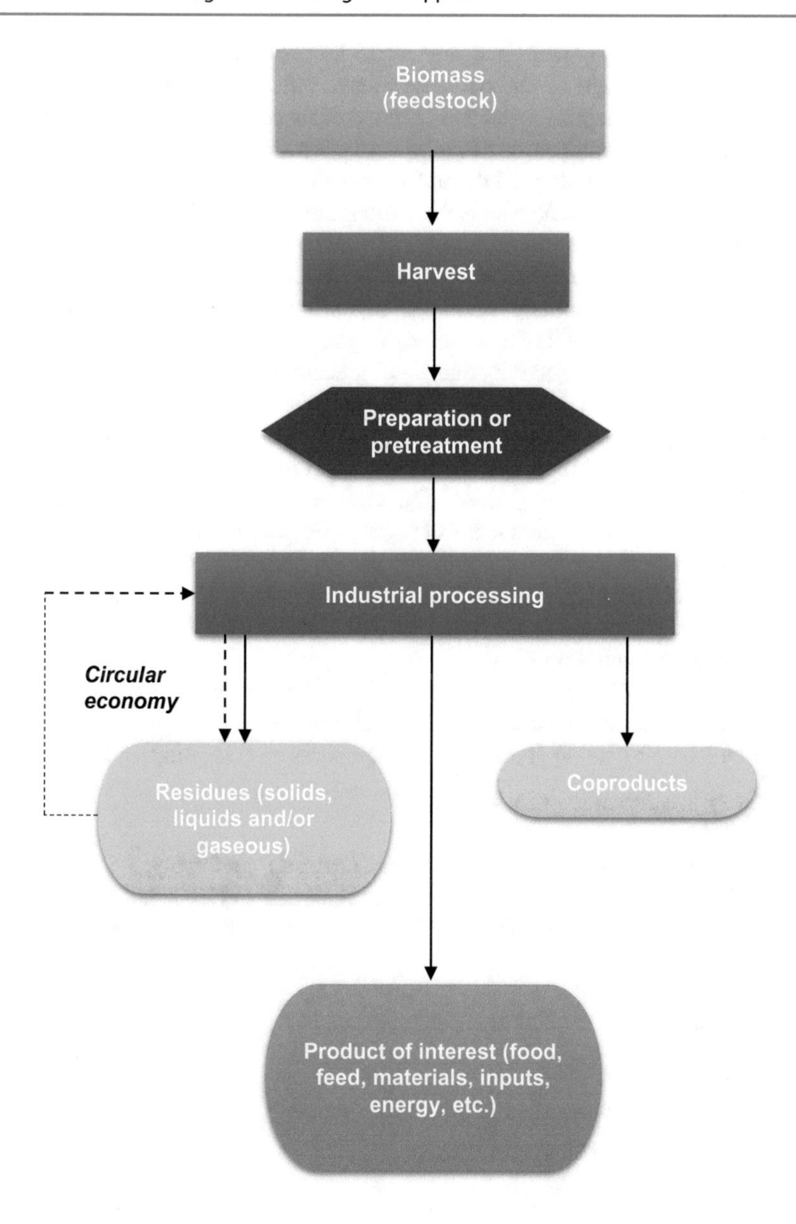

Fig. 5.1 A typical flowchart for the economic exploitation of a certain biomass chain (e.g., grain). The harnessing of residues follows the concept of circular economy (dash). Author

necessary, which will allow the separation of the precursor of interest, and if it does not have the appropriate purity, a purification step is carried out. Once the precursor molecule is obtained, the organic synthesis step starts, in which the search for the best catalysts is involved (catalytic screening of various catalysts: inorganic heterogeneous, and inorganic and enzymatic homogeneous) and the appropriate approach for the design of the synthesis routes. After the synthesis of the target product, it must be properly identified in terms of its chemical structure and purity. Once the chemical identification has been made, it moves on to the step of studying the potential of the target product and its synthesis route, and when they have industrial potential, the next stage is the scaling aiming the industrial production. If the product and route are not economically viable, the search for a new precursor molecule, a new target product or both can be restarted.

It is worth mentioning the development and use of catalysts for these processes, given the importance of them to improve yields and selectivity - considering enantioselectivity, regioselectivity and stereoselectivity. Zeolites have been applied in glycolisation, oxidation, hydrolysis and pyrolysis of carbohydrates and hydrogenation of glycerin [20], and in cracking of lignins [33]. Metals (soluble and insoluble salts, and complexes) have been applied in heterogeneous catalysis (Ni, Pd/C, Ru/C, Co-Mo, Ni-Mo, Ru/Al_2O_3, etc.) for the reduction of lignins and glycerin [5, 33]; metallic complexes of V, Mn, Co, Pd, Fe, Re and Cu, as homogeneous and heterogeneous catalysts for the oxidation of starch and cellulose, among other reactions [7]. Extracted and purified enzymes, on the other hand, such as cellulase, β-glucosidase and xylanase, are widely used in the hydrolysis of cellulose and hemicellulose [23].

This strategy searching for add value to the biomass residues according circular economy (Chap. 3) and bioconomy (Chap. 1) approaches.

A summary for general application of the chemical processing is depicted.

Types of wastes/residues to be treated:

- Lignocellulosic from several sources (dry material), preferably;
- Liquid effluents with volatile organic compounds (VOCs) and semi-volatile organic compounds (SEMIVOCs) by advanced oxidative processes (AOPs) [30].

Advantages:

- Fast kinetic;
- Downstream yield relatively good;
- Generation of high-value products.

Disadvantages:

- CAPEX[1] and OPEX[2] can be very expensive;
- Need of pre-treatment step;
- Can generate secondary residues, as gaseous and liquid effluents.

[1]CAPEX = capital expenditures.
[2]OPEX = operational expenditures.

5.1.2 Biochemical Processing

Biochemical processes are very similar to chemical processes in terms of the characterization of residues or co-products, pre-treatment (when necessary), structural identification and study of industrial potential. However, the main particularities of this type of process concern the use of microorganisms, such as fungi, bacteria, yeasts and microalgae, which have biochemical mechanisms that allow the synthesis of several organic compounds, whether they are building blocks, synthesis intermediates or compounds that have a direct application.

The application of these processes is illustrated in the Fig. 5.3. As for chemical processes, the biomass residue must undergo a complete chemical characterization, which aims to determine its chemical constitution, in addition to physicochemical properties of interest. Then, there is the pre-treatment of the waste, when it is necessary, which will allow the availability of the medium for metabolization by microorganisms. After that, the main step is taken, which is usually fermentation, in which the search for the best microorganisms (screening step) and the appropriate approach for the design of production routes are involved—in some cases it is necessary the application of molecular biology and genetic engineering techniques to enhance the performance of the process. After the biochemical synthesis of the target product, it must be separated from the medium (downstream step) and properly identified as to its chemical structure and purity. Once the chemical identification has been made, the next step is to study the potential of the product obtained and from its biochemical synthesis route, and when it presents industrial potential, the next step is the scale-up aiming the industrial production. If the product and route are not of industrial interest, we can restart the search for a new precursor, a new target product (target compound), or both.

As for the microorganisms used in bioprocesses, the yeast *Saccharomyces cerevisiae* can be highlighted for the fermentation of glucose for the production of ethanol (first or second generation, 1G and 2G) [23]; the bacteria *Euscherichia coli* for the metabolization of glucose and production of compounds as 1, 3-propanediol; *Lactobacillus delbrueckii* for the production of lactic acid via glucose fermentation; and *Anaerobiospirillum succiniciproducens* for the production of succinic acid through the fermentation of sugars (pentoses and hexoses) [5]. However, despite the high potential to treat biomass residues, it is somewhat difficult to consider bioprocesses as substitutes for chemical processes, since the first will hardly reach yields and purities superior than the second, and a synergy between them should be considered—especially when think of the concept of biorefinery.

This strategy also searching for add value to the biomass residues according circular economy (Chap. 3) and bioconomy (Chap. 1) approaches.

A summary for general application of the biochemical processing is depicted.

Types of wastes/residues to be treated:

- Miscellaneous of solid and liquid organic residues with C-, O- and N-derived components.

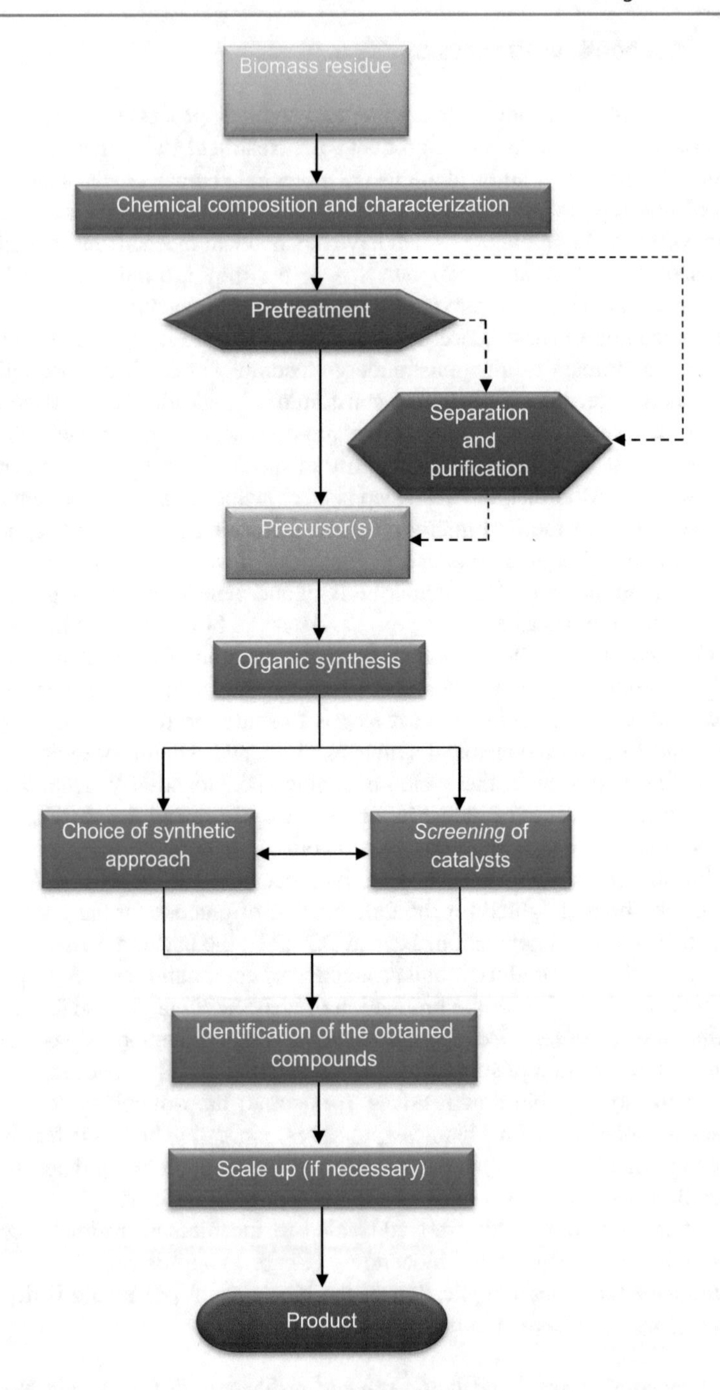

Fig. 5.2 Simplified flowchart of the strategy for the treatment of biomass residues by means the application of chemical processes. It follows the circular and bioeconomy approaches. Author

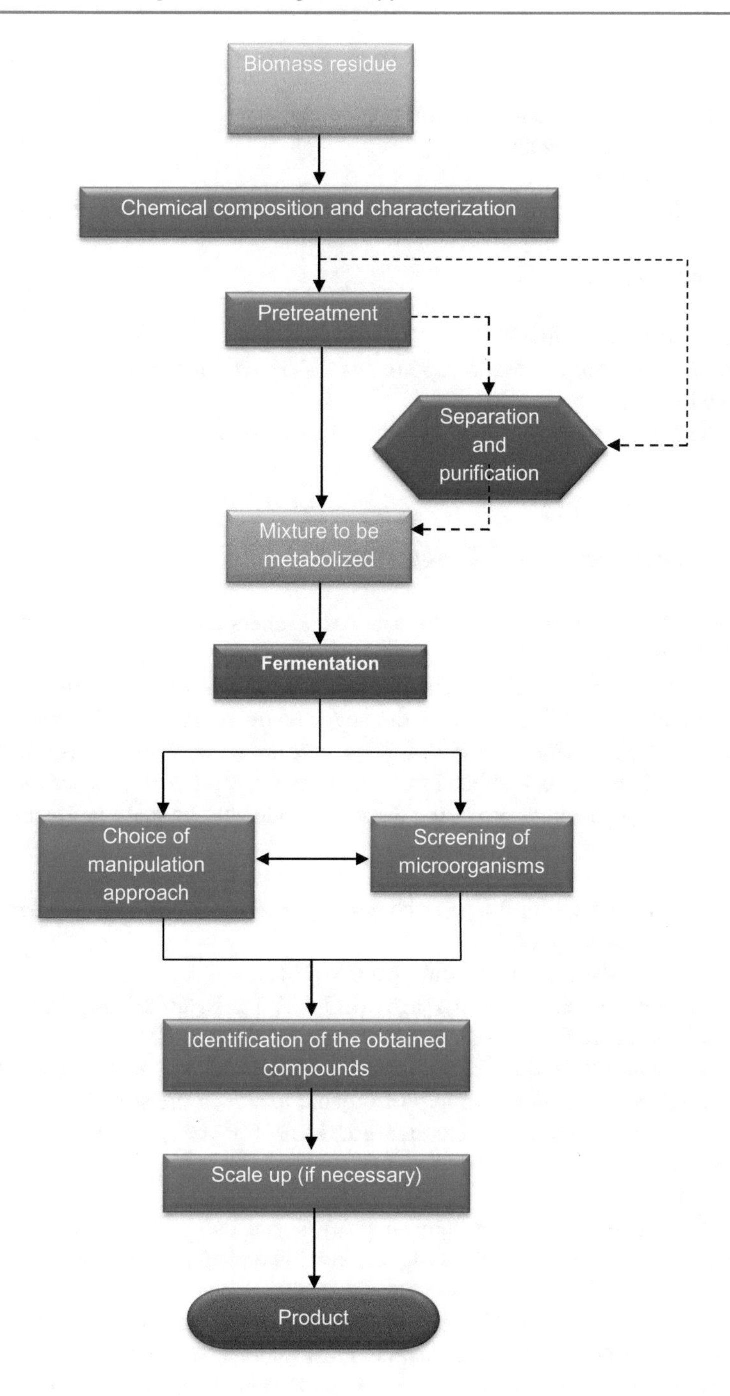

Fig. 5.3 Simplified flowchart of the strategy for the treatment of biomass residues by means the application of biochemical processes. It follows the circular economy and the bioeconomy approaches. Author

Advantages:

- Generally, cheaper than chemical processing;
- Reduced energy demand;
- Ease to operate;
- Common products are bioenergy and biofertilizers, with well-established uses.

Disadvantages:

- Increased water demand;
- Odour release from organic vapours, as S-derived compounds;
- Slow kinetic;
- Limited downstream yield, compromising the generation of high-value products.

5.1.3 Thermochemical Processing

Thermochemical processes can be seen as a chemical processing with a high energetic demand.

As well as chemical and biochemical processes, the steps of waste characterization, structural identification and the study of industrial potential are common. However, the main characteristic of these processes concerns the use of thermal energy, which leads to combustion, carbonization, pyrolysis (fast or slow), torrefaction and gasification, providing different products. The main products of these processes are:

- Carbonization: coal for the production of thermal energy and for metal reduction in the steel industry [22];
- Combustion: thermal and electric energy [20];
- Gasification: synthesis gas or syngas ($CO + H_2$) to be used in organic synthesis of various molecules for use in the chemical industry [2];
- Fast pyrolysis: bio-oil and bio-charcoal (or bio-char), to be used as a substitute for fossil fuels and in the supply of organic matter to the soil (Qu et al. 2011);
- Torrefaction or pre-carbonization: briquettes for the production of thermal energy [9].

Gasification is the thermochemical process that can add greater value to the biomass residue, since from the syngas several chemical compounds of renewable origin are obtained, alternative to petrochemicals.

Figure 5.4 shows the application of thermochemical processing. As for the two previous categories, the biomass residue must undergo a complete chemical characterization step, which aims to determine its chemical constitution, in addition to physicochemical properties that are of interest. Then, there is the thermochemical processing of the residue—there is, in some cases, the need to purify the obtained molecules. After obtaining the target product, it must be duly identified as to its chemical structure and purity. There is the possibility of using a target product as a precursor to other molecules of greater added-value via chemical synthesis, such as the syngas that is used as a reagent in the synthesis of various organic molecules of

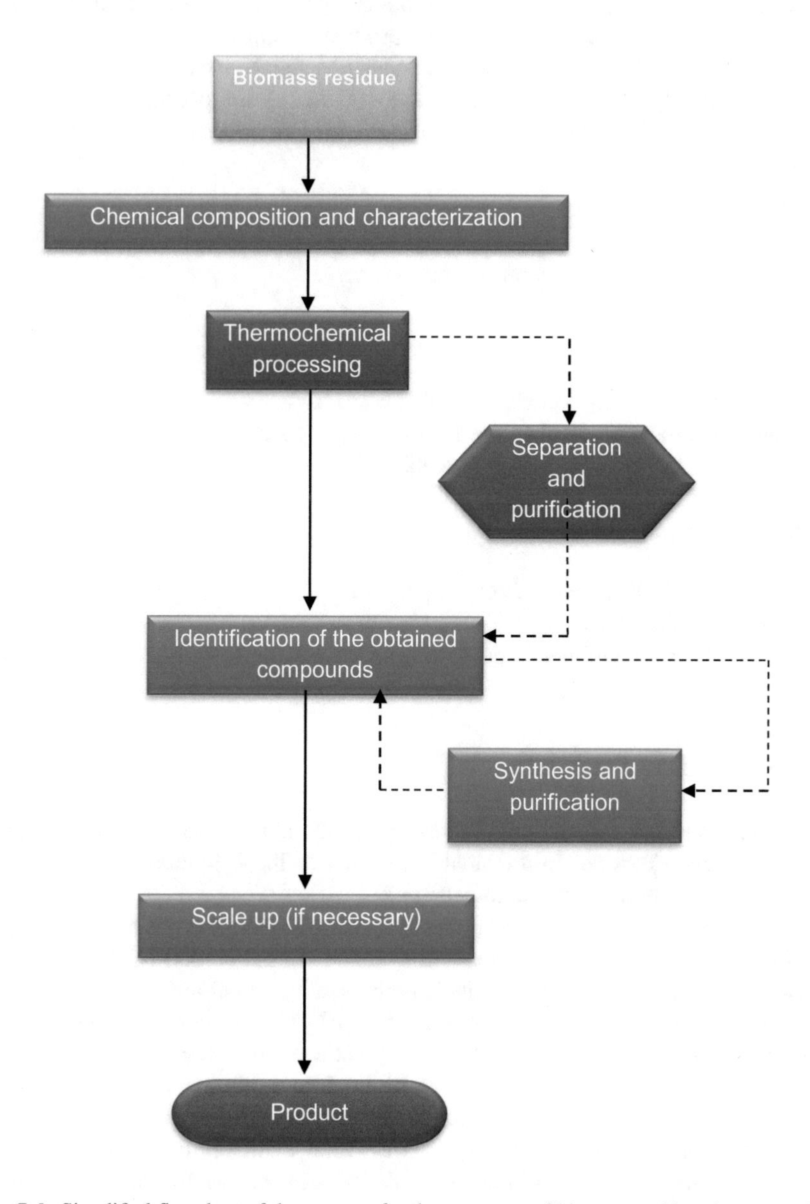

Fig. 5.4 Simplified flowchart of the strategy for the treatment of biomass residues by means the application of thermochemical processes. It follows the circular economy and the bioeconomy approaches. Author

industrial interest, as fuel hydrocarbons, through the Fisher-Tropsh reaction [12]. Once the chemical identification has been made, the next step is to study the potential of the product obtained and the route involved, and when it presents industrial potential, the next step is the scale-up aiming the industrial production. Once again, if the product and route are not of industrial interest, we can restart the search for a new molecule, a new target product, or both.

This strategy also searching for add value to the biomass residues according circular economy (Chap. 3) and bioconomy (Chap. 1) approaches.

A summary for general application of the thermochemical processing is depicted.

Types of wastes/residues to be treated:

- Lignocellulosic from several sources.

Advantages:

- Generally, it has fast kinetic (e.g., fast pyrolysis);
- Common products are charcoal for bioenergy (biochar)—with well-established uses—and bio-oil with a high potential of use.

Disadvantages:

- CAPEX and OPEX can be very expensives;
- High energetic demand, increasing the costs;
- Limited downstream yield, compromising the generation of high-value products.

5.1.4 Physical Processing

We can see in Figs. 5.1, 5.2 and 5.3 that generally is necessary a pretreatment step to turn biomass residues into available material to those processing technologies (chemical, biochemical and thermochemical).

According Virmond et al. [31], physical pretreatments are the most commonly applied for thermochemical biomass conversion. Comminution of lignocellulosic materials is usually carried out through a combination of chipping, grinding, and/or milling. The size of the materials is usually 10–30 mm after chipping and 0.2–2 mm after milling or grinding. The usual practice is that the harvested biomass is naturally dried before being transported by truck to a pretreatment plant. The first step of the pretreatment is size reduction and further active drying, after which the actual pretreatment conversion takes place. Other configurations need to be proposed for other types of agroindustrial residues. Biomass forms which are suitable for cost-effective transport over longer distances and allow transshipment with bulk handling processes are, for instance, chips and pellets. The final particle size and biomass characteristics determine the power requirement for mechanical fragmentation of agricultural materials. The energy consumption for size reduction of hardwoods and agricultural wastes as a function of final particle size and fragmentation ratio (size reduction) was quantified. It was proposed that, if the final particle size is held to the range of 3–6 mm, the energy input for fragmentation can be kept below 30 kWh per ton of biomass.

However, this same methodology can be readequated and applied to another strategies for lignocellulosic residues, as the generation of biogas and biofertilizers or composting.

A summary for general application of the physical processing is depicted.

Types of wastes/residues to be treated:

- Lignocellulosic from several sources (preferable).

Advantages:

- Generally, cheaper than other processing categories.

Disadvantages:

- As it doesn't involve transformation (or conversion), its capacity of treatment and generation of value is limited;
- Association with other(s) processing category(ies) is almost mandatory.

5.2 Application of Monitoring Strategies

Chemical analyses play an important role in the exploitation of biomass, being seen as technologies to support all stages of processing of the crop chains, such as sugarcane, soybeans, corn, forests, paper and cellulose, agroindustrial waste, among many others.

The high heterogeneity and the consequent great chemical complexity of the plant biomass make it the raw material for several final products, such as energy, food, feed, chemicals and materials. Then, biomass residues have also a high complexity that demand chemical analyses, made in the laboratory, in the field or in the agroindustry.

In the study of biomass and its transformation processes, chemical analyses can be applied as follows:

- Determination of the chemical constitution of different types of biomass, their products, co-products, by-products and residues;
- Monitoring of chemical, biochemical or thermochemical transformation processes;

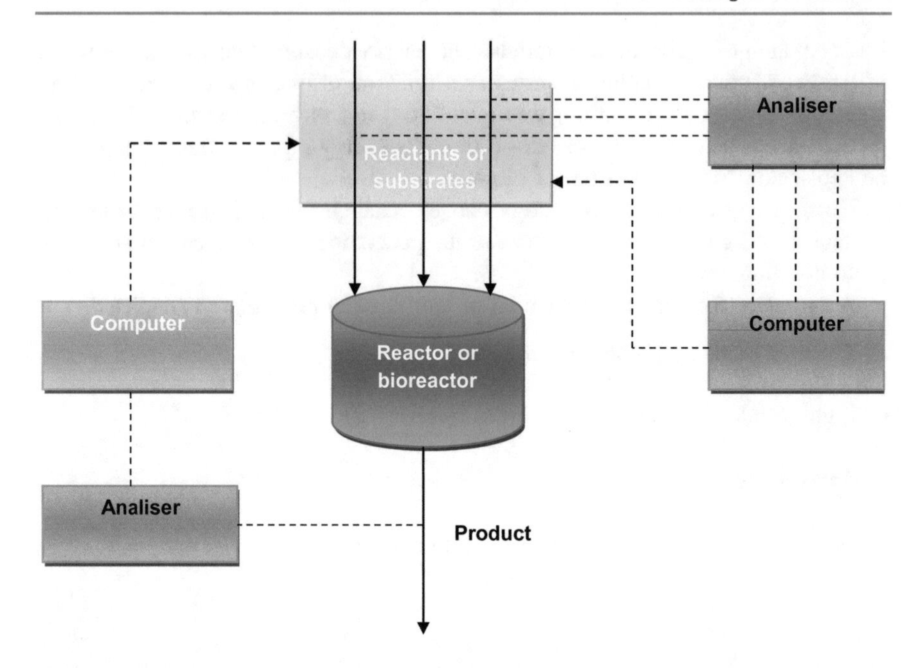

Fig. 5.5 Flowchart of an online analysis system of biomass transformation processes, as a residues fermentation or residues gasification

- Observation of physical and chemical properties and characteristics of biomass and its molecular constituents.

For a fast and reliable analytical data that allow the fast decision taking, the monitoring in real-time is the best strategy. However, it cannot be overlooked in lieu of laboratory analyses for more complex problems, such as the development of new products and processes, critical assessment of the potential use of a new biomass, among other technical and scientific circumstances.

Figure 5.5 describes an online analysis scheme for biomass transformation processes—or treatment process, using mainly spectroscopic probes, such as raman and near and medium infrared.

The flowchart in the Fig. 6.5 will allow the control of process variables in a mode that errors in all steps will be reduced or eliminated.

5.3 Case Studies

The treatment strategies presented in this item are available technologies and available scientific knowledge to turns agroindustry sustainable.

5.3.1 Treatment of Sugarcane Bagasse by Cogeneration

The energy needs of an autonomous distillery, related to the demands for heat, electricity and mechanical energy, are met by the cogeneration plant that consumes the residual biomass (bagasse) generated in the process. Sugarcane bagasse (Fig. 5.6), currently used as fuel in all existing sugarcane plants, is consumed in steam systems that, when operating more efficiently, provide a reduction in fuel consumption and increased generation of surplus electricity (or bioelectricity).

The cogeneration process demands steam at low pressure, usually at 2.5 bar, as a source of heat for processes of treatment and evaporation of the juice and distillation of ethanol. Steam consumption can vary depending on the degree of technology and the existing thermal integration, which directly influences fuel consumption in the boiler. In addition, cogeneration systems that operate with extraction-condensation turbines require low process steam consumption so that the last stage of the turbines (Fig. 5.7) is able to operate with sufficient flow rates to justify the investment.

The systems for preparing the sugarcane and extracting the juice are also consumers of steam to drive the turbines, which provide mechanical energy to the chippers, shredders and milling suits. Normally, steam extraction to meet this demand takes place at 22 bar pressure, and the flow rate varies according to the efficiency of the drive turbine.

Fig. 5.6 A mountain of sugarcane bagasse in a distillery plant in Brazil. Author

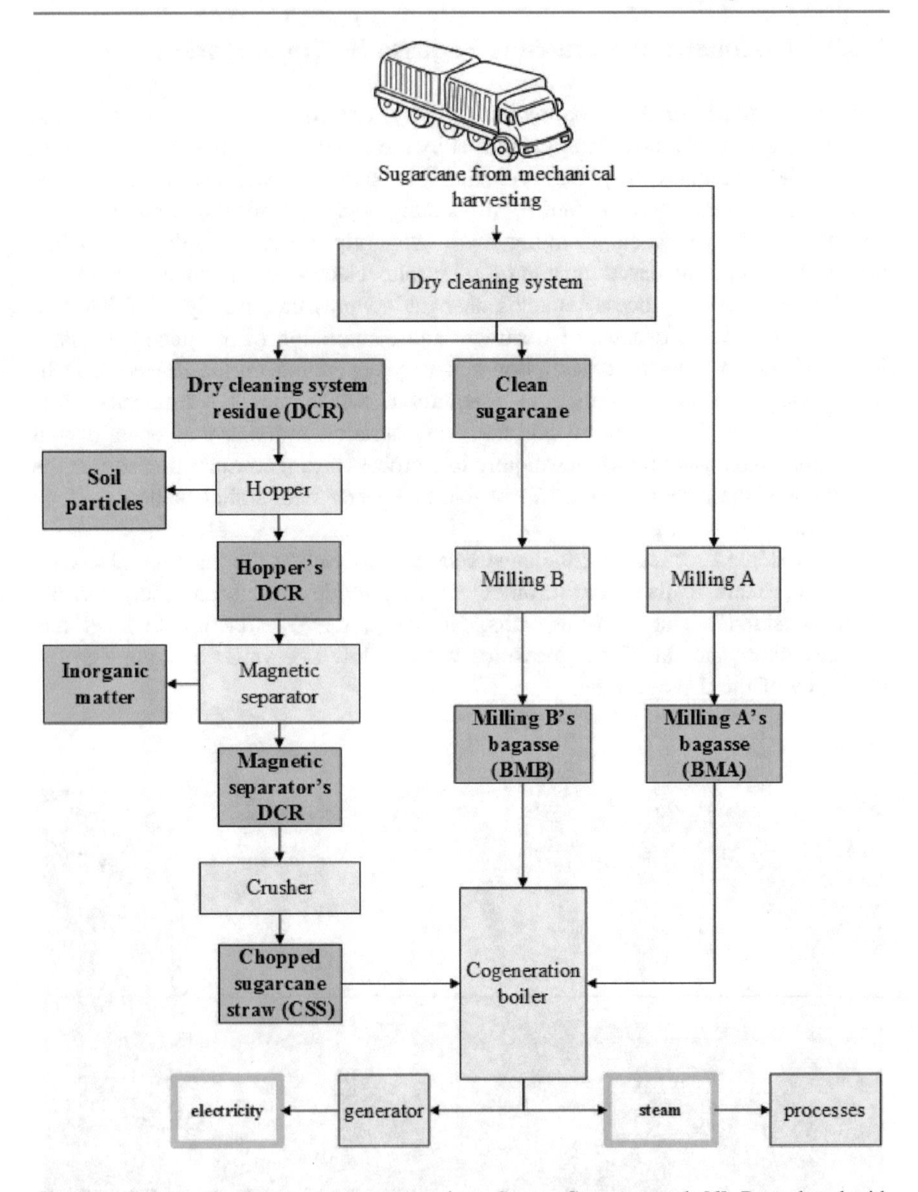

Fig. 5.7 Cogeneration in a sugarcane processing. *Source* Camargo et al. [6]. Reproduced with permission from Elsevier

Another important technological advance regarding to the sugarcane preparation and milling system is the use of electric motors to drive the equipment, replacing steam turbines. The electrification of these systems allows a significant increase in energy conversion efficiency, which results in an increase in surplus electrical generation. Alternatively, some plants have employed hydraulic drive systems, which represent an intermediate solution, from the point of view of efficiency and the generation of electrical surpluses, between the purely electric drive and the purely mechanical drive directly made possible by steam turbines.

Some parameters adopted in the procedure for calculating the cogeneration process settings are shown below:

- Process steam pressure: 2.5 bar
- Process steam temperature: 128 °C
- Deaerator operating pressure: 1.3 bar
- Boiler efficiency: 85%
- Isentropic efficiency of the steam turbine: 78%
- Isentropic efficiency of pumps: 65%
- Return condensate temperature: 79 °C
- Loss of steam in the process: 5% of the process steam
- Steam consumed in the deaerator: 5% of the live steam generated
- Lower calorific value of bagasse: 7524 kJ kg^{-1} (50% humidity)
- Lower calorific value of straw: 12,960 kJ kg^{-1} (15% humidity)
- Lower calorific value of lignin: 16,219 kJ kg^{-1} (50% humidity)
- Total bagasse production: 140 kg t^{-1} cane (dry basis)
- Total straw production: 140 kg t^{-1} (dry basis)
- Lignin production: 200 kg t^{-1} hydrolyzed bagasse (dry basis)

The cogeneration inside the sugarcane processing is illustrated in the Fig. 5.7. Figure 5.8 depict a sugarcane processing plant.

Fig. 5.8 A sugarcane processing plant in the Brazilian southeastern region. This plant generates bioelectricity from cogeneration. Author

5.3.2 Treatment of Pig Slurry by Fermentation to Generate Biogas

This case study was based on Do Amaral et al. [8], which describe the behavior of an up flow anaerobic digester and the increasing in the organic loading rate.

Experimental design

The experiments were conducted at the experimental unit of Embrapa Swine and Poultry, in Concórdia, Santa Catarina State, Brazil. The swine effluent used in this study came from two farrow-to-finish pig farms with capacity for 3800 animals. The effluent was stored in a gutter collection system; subsequently, it was sent by gravity to the swine manure treatment plant (SMTP), passing through a flow damping box with adjustable gates and a 2 mm rotating sieve for coarse solid retention, and then homogenized in the equalization tank (ET). In this study, the effluent was directly repressed from ET according to the required flow. A BioKöhler® digester was used, which was produced in fiberglass, with an upflow system and working volume of 10 m^3. The internal temperature (36 ± 2 °C) was maintained by using a serpentine system for water recirculation throughout the experimental period. Biomass was agitated by recirculating the effluent itself for 2 min every day to prevent clogging.

Figure 5.9 shows a schematic biogas plant.

Loading progression in volatile solids

It was performed in three stages as shown in Table 5.1.

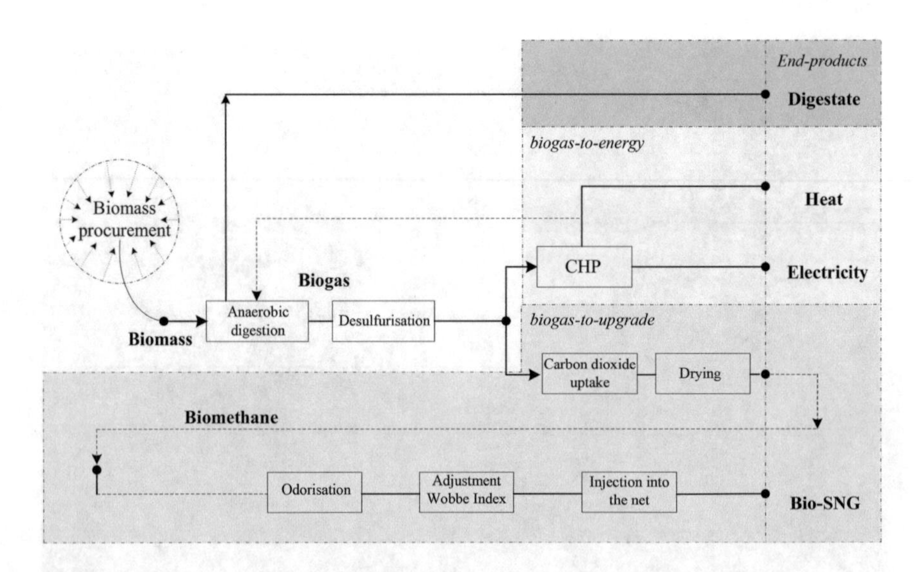

Fig. 5.9 A schematic biogas plant highlighting the key valorisation steps and pathways. SNG = synthetic natural gas. *Source* Weinand et al. [32]. Reproduced with permission from Elsevier

Table 5.1 Description of the studied stages with respective loading flow rates, hydraulic retention time (HRT) and estimated loading of volatile solids (VS), and chemical oxygen demand (COD)

Stage	Flow (m^3d^{-1})	HRT[a] (d)	Estimated loading	
			kgVS m^{-3} d^{-1}	kg O_2 m^{-3} d^{-1}
1	0.56±0.01	17.86±0.10	0.45±0.02	1.16±0.01
2	1.32±0.01	7.57±0.06	1.00±0.01	2.74±0.02
3	1.88±0.01	5.32±0.03	1.50±0.01	3.90±0.02

Source Do Amaral et al. [8]. Reproduced with permission from corresponding author

Digester loading was intermittently performed, due to the minimum work flow of the repression system, via a submerged pump in ET controlled by a timer. Considering the preestablished daily flow for each stage, the loading pump was activated for 3.2 min every two hours during stage 1; 3.2 min every 1 h during stage 2; and 4.8 min every 1 h during stage 3.

Sampling

Samples were collected at two locations (input and output), representing the swine effluent before and after the anaerobic treatment, and then stored at 4 °C until analysis. Sample collections were carried out in different periods from the agitation system operation.

Analytical determination

Chemical oxygen demand (COD): COD analyses were performed based on the sample acid digestion in the presence of potassium dichromate in a closed reflux system, held in a digester at 150 °C for 2 h. After sample cooling, absorbance reading was carried out in a spectrophotometer at 620 nm.

Total solids (TS) and volatile solids (VS): samples were dried at 105 °C, until constant weight, for determination of the TS level, and then muffle furnace calcined at 550 °C for 1 h for determination of the fixed solids (FS). The VS content was obtained by the difference between TS and FS. *Biogas production:* it was determined by a gas meter.

Biogas composition: determination of CO_2 was performed by adapting the Orsat method: NaOH solution reacts with CO_2, absorbing it in solution as carbonate and bicarbonate. The difference between initial and final volumes was used to estimate CO_2 and methane concentration; H_2S was determined by the method of methylene blue, reacting with ferric sulfide and dimethyl-*p*-phenylenediamine in acid to produce methylene blue. After the reaction, ammonium phosphate was added to eliminate the color produced by excess ferric chloride. Concentration was then determined by colorimetric comparison.

Statistical analysis: it was performed using GraphPad Prism software[3], version 3.02. A significance test among means was performed at 5% confidence level ($P < 0.05$); the *n* value varied for each studied item.

[3]https://www.graphpad.com/scientific-software/prism/.

5.4 Conclusions

The increase of organic loading in the digester presented a high maximum biogas generation capacity (MBGC) per added volatile solid. This effect is enhanced by the HRT decrease, as MBGC was six times greater when the volumetric organic load was increased from 0.436 to 1.853 $kg_{VS}\ m^{-3}\ d^{-1}$ with decreasing HRT from 17.86 to 5.32 days. This is highly significant for the proposed system when the objective is to produce biogas. However, the COD decrease had an opposite effect to the MBGC, that is, the digester mineralization capacity was reduced according to lower HRT, what affects effluent quality. To minimize this effect, a pre-hydrolysis stage may be added to the system, which may contribute to increased substrate availability to methane-producing microorganisms, greater MBGC, and COD reduction of the final effluent.

5.4.1 Treatment of Black Liquor from Pulp and Paper by Coagulation and Acid Precipitation

This case study was based on Garg et al. [11], which describe the application and the effectiveness of coagulation (using aluminium-based chemicals and ferrous sulfate) and acid precipitation (using H_2SO_4) processes for the pre-treatment of diluted black liquor obtained from a pulp and paper mill from the chemical processing of wood (e.g., kraft process).

Figure 5.10 shows a pulp and paper industry.

Fig. 5.10 A pulp and paper processing plant in Brazil. Courtesy of Suzano

According Zaied and Bellakhal [34], black liquor is one of the main by-product of pulp paper industry, which is considered as pollutant because it contains about 50% m/m of lignin. In addition to lignin the black liquor contains aliphatic acids, acid greases, resins and polysaccharides. This organic matter, mainly dissolved, creates a chemical and biochemical oxygen demands—COD and BOD—some relatively high, and the rejection of this effluent in nature without any treatment is responsible of serious damage for the environment.

Experimental procedure

The experimental runs were performed in a three-necked glass reactor (GR) having a capacity of 500 mL. A vertical glass condenser was attached to the central neck of the reactor to prevent any loss of vapour to the atmosphere during the coagulation/acid precipitation at temperatures higher than ambient temperature. Thermo-well fitted to one of the side necks was used to monitor the temperature of the mixture during the run. The third neck was kept closed using a glass stopper and used only for the withdrawal of samples during experimental run. The original pH value of the wastewater was adjusted to the initial desired value by adding H_2SO_4 before transferring the wastewater to the GR. A pre-determined dose of coagulant was added to the reactor. The experimental runs were performed at temperatures ranging from ambient to 95 °C (i.e., 25, 40, 60, 70, 80 and 95 °C). The reactor contents were heated on a hot plate to achieve the desired temperature. The hot plate was equipped with a proportional integral derivative (PID) temperature controller that was connected with a thermocouple (placed in the thermo-well). It took about 30 min to attain the wastewater temperature from ambient to 95 °C. The time at which reaction mixture attained the desired temperature was taken as 'zero time' and at this point, the agitation of the wastewater–coagulant mixture was started with a magnetic stirrer. It must, however, be stated that during the heating of the reaction mixture up to the desired temperature, coagulation–flocculation did take place and COD got removed. Rapid agitation of the reaction mixture for 5 min was followed by slow agitation for 30 min. The mixture was then allowed to cool for 1 h. The supernatant was used for the determination of the COD and colour of the treated sample. The final pH value of the solution after the reaction was also recorded. The cooled reaction mixture was again stirred for uniformity and the slurry so formed was used to study the settling characteristics of the flocs/solids. The settling study was conducted in a 100-mL graduated cylinder. Effect of temperature (during coagulation process) on the settling characteristics was observed and critical solid concentration was calculated using sedimentation data.

Apart from this, the metal ion concentration in the supernatant and the sludge was also examined. The solid residue obtained after the treatment was subjected to thermal analysis to determine its high heating value (HHV).

Analytical methods

COD of the effluent was determined by using standard dichromate open reflux method. Colour of the untreated and treated wastewater was measured at a wavelength of 363 nm using an UV/VIS spectrophotometer. The concentration of metal ions in the wastewater and the sludge was determined using an atomic absorption spectrophotometer.

Elemental analysis for C, H, N and S was performed using a CHNS analyzer. The HHV of the precipitated dried sludge was determined by preparing a suitable pellet and combusting it in a standard adiabatic bomb calorimeter.

5.5 Conclusions

All aluminium-based coagulants and ferrous sulfate were found to be equally efficient in removing COD of the diluted black liquor (*ca.* 60% reduction for a substrate having an initial COD = 7000 mg L^{-1}). Commercial alum was found to be the most economical coagulant among these tested coagulants. The capability of a coagulant for reducing COD and colour strongly depends on the pH of the wastewater. Using commercial alum, the maximum COD and colour reductions of 63% and 90% were obtained at an optimum pH value of 5.0. The pH value of the resulting effluent after treatment with alum was decreased. Acid precipitation using H_2SO_4 exhibited substantial COD reduction at a pH value of <4.0. The COD removal was mainly due to the separation of non-readily biodegradable lignin from the wastewater. Further COD reduction can be achieved by treating the supernatant using biological methods or wet oxidation processes or adsorption.

Settling characteristics of the resulting slurry were found to have improved with an increase in treatment temperature. The settling rate of the wastewater treated at 95 °C was almost 3 times of that treated at 25 °C. The solid residue obtained after coagulation and acid precipitation was found to have considerably high heating value and can be used as a renewable source of energy in several energy intensive industries for electricity production and heat recovery. It can be concluded that this kind of fuel will be in demand because of low or negligible cost, biogenic fraction, better thermal characteristics and continuous depletion in coal reserves.

5.5.1 Treatment of Wastes from Agriculture and Food by Advanced on-Site Composting

Composting is the biochemical process of decomposition and recycling of organic matter contained in animal or vegetable wastes to form a compound—not to be confused with a chemical compound. Composting provides a useful destination for organic waste, avoiding its accumulation in landfills and improving soil structure. This process allows dispose of agricultural, industrial and domestic organic waste, such as food scraps and garden waste. It results in a product—the organic compound—that can be applied to the soil to improve it without causing risks to the environment.

Figure 5.11 shows an outdoor composting system.

Fig. 5.11 Outdoor composting system for agricultural and food residues in Brazil. *Source* Barreira [3]. Reproduced with permission from the University of São Paulo

Composting steps

The industrial composting process include three main steps:

i. Separate and select the raw material (waste);
ii. Analyses: physical (real and apparent density, particle size and content contaminants) and chemical (micro and macronutrients and toxic metals);
iii. Statistical analysis for quality control and production of final formulations (compost).

According Francisco Neto [11], the composting process occurs for a diverse population of microorganisms and necessarily involves two distinct phases, the first being active degradation (necessarily thermophilic) and the second maturing, also called curing. In the active degradation, the temperature must be controlled to thermophilic values, in the range of 450–650 °C. In the maturation phase, in which the organic matter previously stabilized in the first phase takes place, the process temperature must remain in the mesophilic range, that is, lower that 450 °C. During the entire process, heat is produced and released, mainly carbon dioxide and water vapor.

During the composting process, many of the components of organic matter are used by the microorganisms themselves for the formation of their tissues, others are volatilized and are biologically transformed into a dark substance with uniform and amorphous mass aspects and have physical and chemical properties different from the original raw material [4]. It is the humic substances (as seen in the Chap. 3).

The compounds produced must have high quality to be considered as soil conditioners. Composting plants must have a structure compatible with the volume generated locally and apply knowledge multidisciplinary teams to monitor the factors that govern composting in the courtyard.

5.5.2 Treatment of Wood Chips by Pelletizing for Energy

According Whittaker and Shield [29], wood pellets are made from dried biomass that has been finely powdered and passed through a pelletizing device at high temperature and pressure. The act of compression and the high temperature causes the lignin within the biomass to melt, gluing the particles together, which then re-form as a solid pellet after cooling.

Mashed et al. [18] stated that the pellet manufacturing process is dependent on the raw material used but tends to include the following steps:

(a) Reception of raw material
(b) Screening
(c) Grinding
(d) Drying
(e) Pelletizing
(f) Cooling
(g) Sifting
(h) Packaging

The resulting product is a high-value, high-density pellet that is consistently shaped and thus more efficiently transported compared to woodchips. Wood pellets have a bulk density of up to 750 kg m^{-3} and a lower heating value of 16.5 GJ ton^{-1}. As a comparison, woodchips have a bulk density of about 250 kg m^{-3} and energy content of 13 GJ ton^{-1}.

The pelletizing process is illustrated in the Fig. 5.12.

Figure 5.13 shows a pelletizing processing plant.

5.5.3 Treatment of Orange Peels by Means Physicochemical Extraction

This case study is based on Lachos-Perez et al. [16], which described a two-step hydrothermal process using subcritical water for sequential removal flavanones and sugars from orange peel for industrial purposes. Citrus peels are an especially relevant example of an under-utilized waste resource due to their composition and volume produced every year by the fruit juice industry.

Figure 5.14 shows a processing plant of orange juice.

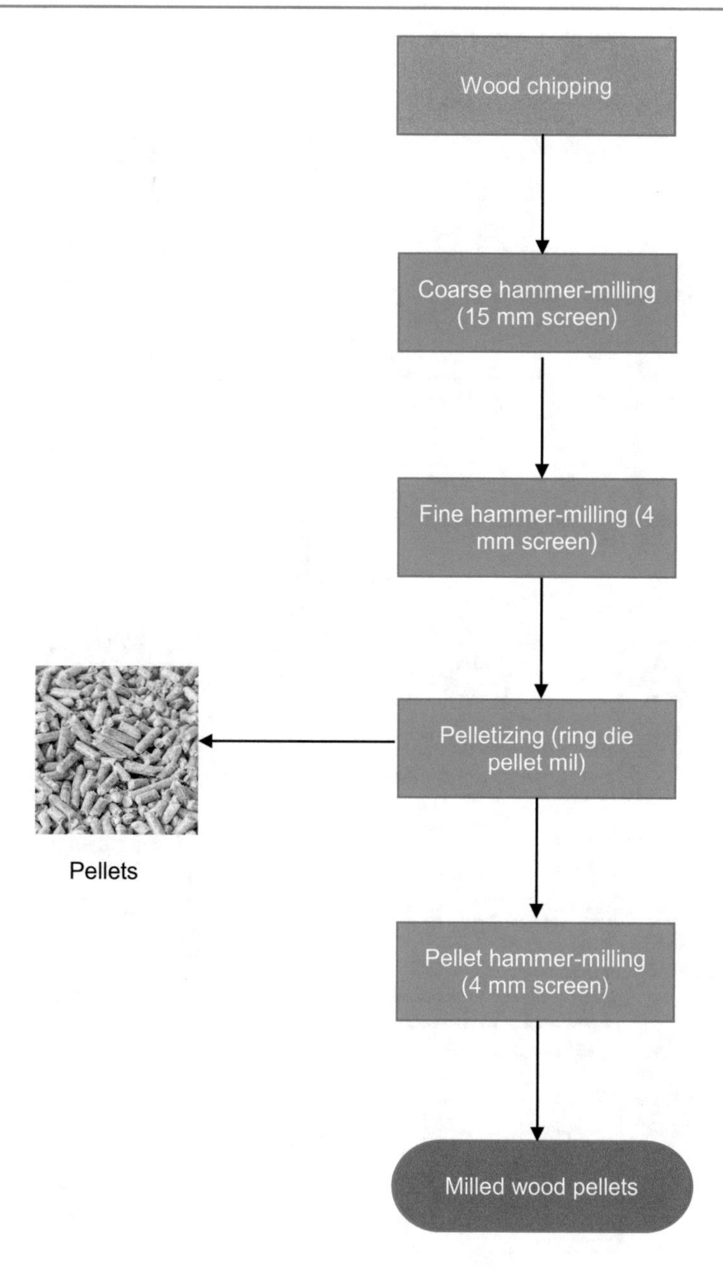

Fig. 5.12 Flowchart of the pelletizing process. *Source* adapted from Masche et al. [18]. Reproduced with permission from Elsevier

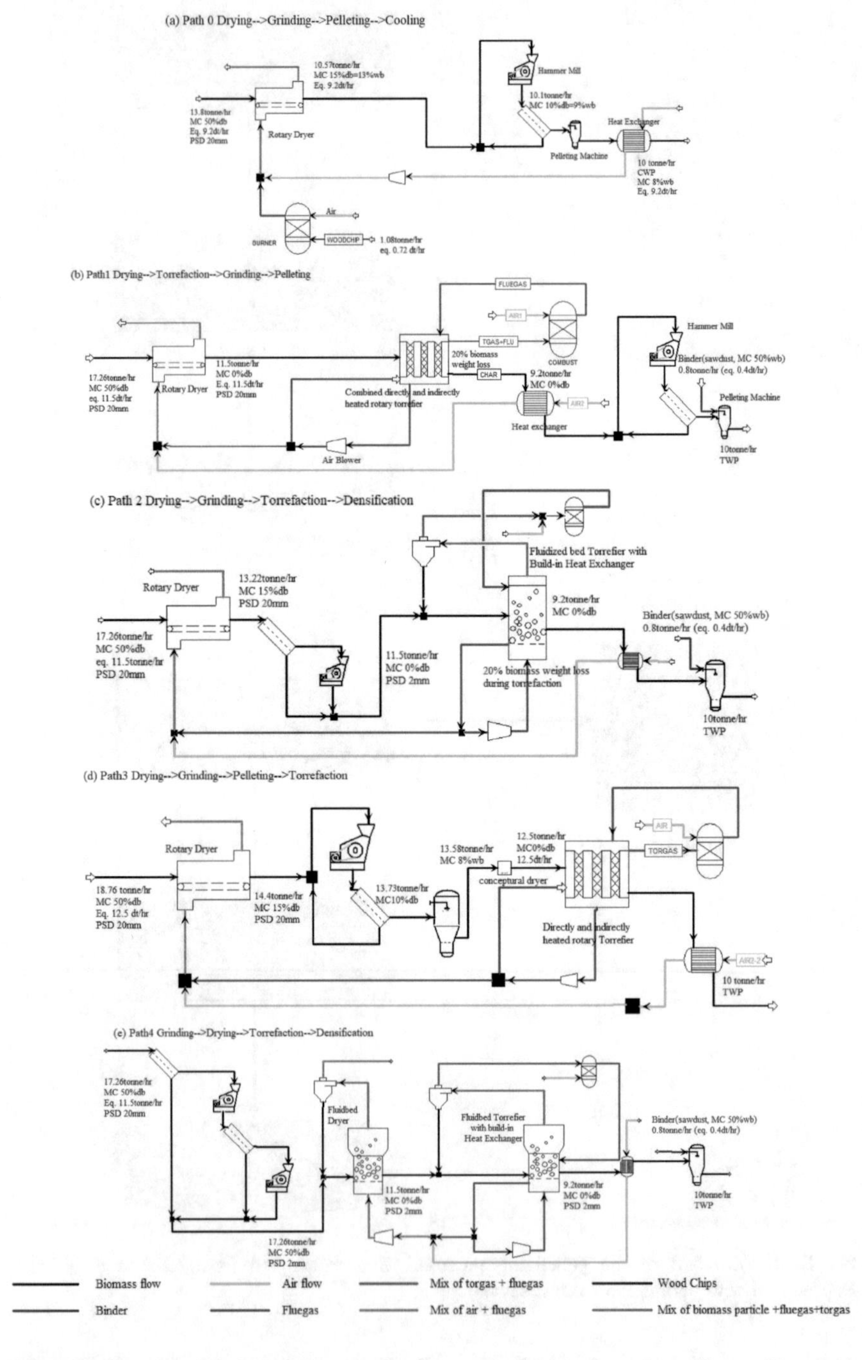

Fig. 5.13 Conceptual design of five wood pellet production configurations *Source* Yun et al. [35]. Reproduced with permission from Elsevier

Fig. 5.14 Processing plant of orange juice in Brazil. Courtesy of Cutrale

Sequential subcritical water process

Orange peel (OP) extraction-fractionation experiments were conducted in a semi-continuous flow reactor (i.e., batch for the solid and continuous for the liquid) and following 2 steps.

In the first step, 5.0 g of OP were loaded into the reactor. Subcritical water extraction conditions were chosen in accordance with results obtained in previous work, in which the best condition (temperature: 150 °C) to recover two distinct types of bioactive compounds (flavanones: hesperidin and narirutin) using subcritical water was identified. In the second step of experiments, sequentially after recovering flavanone-rich extracts, the reactor was then heated to three different temperatures: 200, 225, and then 250 °C. The pressure was kept constant at 100 bar by introducing pressurized water into the system and water flow rates were evaluated at 10, 20, and 30 mL min^{-1} for both experiments procedures (extraction and fractionation).

The liquid product was collected after reaction and analyzed to quantify concentrations of sugars, bioactive compounds, antioxidant activity, and several major degradation products. The liquors were filtered and stored at 4 °C prior to analysis. The solids remaining in the reactor after fractionation were collected in glass flasks, weighed, and stored at −18 °C. The collected biomass was dried at 105 °C for 24 h. Solid products were analyzed using FTIR spectroscopy and thermogravimetric analysis.

Analytical methodology

Reducing sugars (RS) content of the hydrolyzate was determined using colorimetric method. The samples were subjected to acid hydrolysis conditions to decompose sugar oligomers into monomers prior to detection as total reducing sugars (TRS). After the coloring reaction, sample absorbance at 540 nm was measured using a spectrophotometer.

Total phenolic content was determined using the Folin–Ciocalteau method. Each extract was diluted in distilled water. Triplicates of 0.5 mL diluted sample and 0.5 mL Folin–Ciocalteu reagent were mixed and incubated at dark at room temperature for 3 min. Then, 0.5 mL of saturated sodium carbonate solution and 3.5 mL distilled water were added and the mixture was again stored in the dark for 2 h at room temperature. The absorbance was recorded at 725 nm using a spectrophotometer. A gallic acid standard curve was determined at 725 nm and samples were quantitated as gallic acid equivalents (GAE).

The antioxidant capacity of the extracts obtained from OP was determined by spectrophotometry using the 2,2-diphenyl-1-picrylhydrazyl (DPPH) scavenging assay, ferric reducing antioxidant power (FRAP), and oxygen radical absorbance capacity (ORAC).

Sugar monomers were analyzed by HPLC-PAD to analyze released monosaccharides (glucose and xylose) by anion exchange chromatography.

Analysis of flavanones was performed by by HPLC-UV. Inhibitor concentrations present in the hydrolyzate were also measured by HPLC-UV.

5.6 Conclusions

The highest yields of hesperidin and narirutin (22.9 ± 0.7 and 1.9 ± 0.2 mg g^{-1} OP, respectively) were obtained at 150 °C and 10 mL min^{-1}. Extraction times and S/F (solvent/feed) ratios need to be optimized during the dynamic process. Optimal sugars yields were observed at 200 °C, reaching values of 7.14 ± 0.95% and 13.44 ± 1.00% of arabinose and glucose, respectively, which indicates economic potential for subcritical water fractionation of OP. Solid residue characterization provided further information about pectin extraction, hemicellulose hydrolysis, and char formation during treatment that can guide future development efforts.

5.7 Environmental Management

As previously mentioned, the agroindustry and food sectors produce large amounts of waste, both liquid as solids. These residues can present high problems of final disposal and potential pollutant activity, in addition to often representing losses of biomass and high-value nutrients. To contrary to what happened in the past, when waste was disposed of in landfills or employees without treatment for animal feed or fertilizer, currently, minimization concepts, recovery, utilization of by-products and waste bioconversion are increasingly widespread and necessary for agroindustrial chains [17].

Agroindustrial waste/residue is generated in the processing of food, fibers, leather, wood, sugar and alcohol production, etc., and their production is generally seasonal, conditioned by the maturity of the culture or supply of the raw material. Wastewater can be the result product washing, scalding, cooking, pasteurization, cooling and washing of the product processing equipment and facilities. Solid waste consists of leftovers process, waste and packaging waste, sludge from water treatment systems residues, in addition to garbage generated in the cafeteria, patio and agroindustrial office [19].

Generally, all consumer goods are potential solid waste. Everything that is produced by human activity and consumed in homes, businesses and industries, after having no more use, can be separated, selected and processed, resulting in waste or residues, as those from agriculture and agroindustry. These residues must have their correct treatment and fate. For waste management, all types of treatment residues can be enclosed into these categories:

- Recycling: consists of the reintroduction of waste in the product process. It generates energy savings in the production processes, and reduces the use of raw materials. It is a strategy that uses physical processing before the chemical processing.
- Biodigesting: It consists of the decomposition of organic matter in the absence of oxygen in the biodigesters, generation the biogas (mostly, methane). It is a strategy that uses biochemical processing.
- Composting: carried out through a biochemical process of decomposition of organic matter, the final result is an organic compound, which can be used in the soil as biofertilizer without causing risks to the environment; widely used in the countryside, but in the urban environment, sorting and ridding the organic component of toxic or dangerous components must be carried out.
- Landfill: form of final disposal of solid waste on the ground, in a strategic location and with engineering techniques, following operational rules to avoid damage to public health and environmental impacts.
- Incineration: a process that consists of reducing the weight and volume of waste by controlled combustion. This method is even more used for the treatment of hospital and industrial waste. It is a strategy that uses thermochemical processing.

The certification in ISO 14001 norm, Environmental Management System [14] help in the feasibility studies of monitoring and improvements in solid waste treatment projects and in the company's environmental policy in general. This norm provides a guideline to keep the organization within the laws regarding the company's field of action, offering an efficient environmental management system and, consequently, the solid waste management.

Schenini [24] proposed a set of operational and management actions to be applied in the environmental management of agroindustries, taking as reference the previous cited ISO norm adapted to this industrial environment. These actions are:

(a) Operational actions:

- Mass balance: identification and quantification of infrastructure resources as energy, water, industrial gases and fuels.
- Anticipation and monitoring: adoption of measures to prevent ecological accidents or damage to nature, using systematic fixed and mobile monitoring and operational support with containment basins, "lung" tanks, emergency dikes and preparation of brigades to combat claims.
- Ecological product and LCA: the analysis of the life cycle of the products allows to know a product since its conception, its raw materials, its processes, residues and losses, until its disposal and reuse (e.g., cradle-to-grave approach).
- Cleaner production: an integrated view of company, seeking to eliminate or minimize aspects and impacts, in addition to decrease in consumption of raw materials and infrastructure.
- Reverse logistics: it is basically concerned with collecting, packaging, transport and dispose of the residues that were generated in the activities of obtaining the raw material, in production, commercial distribution and post-use.

(b) Management actions:

- Adoption of sustainable strategies and policies as a basis for the environmental management system (EMS).
- Environmental management system (EMS): as an administrative and managerial mechanism to monitor the organization's performance. It includes the organizational structure, planning activities, responsibilities, practices, procedures, processes and resources to develop, implement, conclude, review and maintain the environmental policy.
- Environmental audit: to prove the achievements, the audit it needs evidence and this is achieved by making the records and storing documents that prove the achievements.
- Environmental education: to guarantee the effectiveness of the sustainable actions carried out, changes in the organizational culture are necessary, an effect that is achieved when they are implemented. Educational actions such as sensitization, and training in environmental management activities and themes.

Those treatment strategies previously presented in the item 6.1 should be within the environmental management vision to reach the best results, specially the sustainability of products and processes in the agroindustry.

5.8 Conclusions

Agricultural biomass wastes/residues are predominantly crop stalks, leaves, roots, fruit peels and seed/nut shells that are normally discarded or burned but are in practice a potential valuable supply of raw material. However, there are some challenges in trying to determine the extent of crop-produced biomass in relation to what is a 'loss' (from production, post harvesting and processing), or a 'waste' (retail or consumer loss). Then, the choice of a treatment strategy is not a simple task; in some cases will be necessary an association of processes—chemical, biochemical, thermochemical and physical categories—to achieve the better result. For instance, the physical processing associated to the thermochemical processing.

The application of treatment strategies using cogeneration (or combustion) for sugarcane bagasse, fermentation for pig slurry, coagulation and acid precipitation for black liquor, advanced on-site composting for brewing waste, pelletizing for wood chips, and subcritical water extraction for orange peel are very good examples of available technologies and scientific knowledge for the agroindustry to turns it sustainable. Furthermore, these technologies and knowledge are better explored when used within an environmental management vision.

References

1. American Chemical Society (2020) Green chemistry. https://www.acs.org/content/acs/en/greenchemistry.html. Accessed Apr 2020
2. Akay G, Jordan CA (2011) Gasification of fuel cane bagasse in a downdraft gasifier: influence of lignocellulosic composition and fuel particle size on syngas composition and yield. Energy Fuels 25:2274–2283
3. Barreira LP (2005) Avaliação das usinas de compostagem do Estado de São Paulo em função da qualidade dos compostos e processos de produção (Evaluation of composting plants in the State of São Paulo according to the quality of the compounds and production processes). Doctoral thesis. University of São Paulo, São Paulo, Brazil
4. Bidone FRA, Povinelli J (1999) Conceitos básicos de resíduos sólidos (Basic concepts of solid residues). University of São Paulo, São Carlos, Brazil
5. Bozell JJ, Petersen GR (2010) Technology development for the production of biobased products from biorefinery carbohydrates—the US Department of Energy's Top 10 revisited. Green Chem 12:539–554
6. Camargo JMO, Gallego-Rios J, Neto AMP, Antonio GC, Modesto M, Toneli JTCL (2020) Characterization of sugarcane straw and bagasse from dry cleaning system of sugarcane for cogeneration system. Renewable Energy, in press
7. Collinson SR, Thielemans W (2010) New materials focusing on starch, cellulose and lignin. Coordination Chem Rev 254:1854–1870

8. Do Amaral AC, Kunz A, Steinmetz RLR, Contelli F, Scussiato LA, Justi KC (2014) Swine effluent treatment using anaerobic digestion at different loading rates. Engenharia Agrícola 34:567–576
9. Felfli FEF, Luengo CA, Soler PB (2003) Torrefação de biomassa: características, aplicações e perspectivas (Biomass torrefaction: characteristics, applications and perspectives). In: Encontro de Energia no Meio Rural (Energy Meeting in Rural Areas). http://www.proceedings.scielo.br/scielo.php?script=sci_arttext&pid=MSC0000000022000000200003&lng=en&nrm=abn. Accessed Apr 2020
10. Food and Agriculture Organization of the United Nations (2011) Global forest products facts and figures. http://www.fao.org/fileadmin/user_upload/newsroom/docs/2011%20GFP%20Facts%20and%20Figures.pdf. Accessed Apr 2020
11. Francisco Neto J (1995) Manual de horticultura ecológica: guia de auto-suficiência em pequenos espaços (Ecological horticulture manual: guide to self-sufficiency in small spaces). Nobel, São Paulo, Brazil
12. Garg A, Mishra IM, Chand S (2010) Effectiveness of coagulation and acid precipitation processes for the pre-treatment of diluted black liquor. J Hazardous Mater 180:158–164
13. Gökalp I, Lebas E (2004) Alternative fuels for industrial gas turbines (AFTUR). Appl Thermal Eng 24:1655–1663
14. International Organization for Standardization. ISO 14001:2015 Environmental management systems. ISO, Geneva
15. Kamm B, Gruber PR, Kamm M (2006) Biorefineries: industrial processes and products: status quo and future directions. Wiley-VCH, Weinheim
16. Lachos-Perez D, Baseggio AM, Torres-Mayanga PC, Ávila PF, Tompett GA, Marostica M, Goldbeck R, Timko MT, Rostagno M, Martinez J, Forster-Carneiro T (2020) Sequential subcritical water process applied to orange peel for the recovery flavanones and sugars. J Supercritical Fluids 160:104789
17. Launfenberg G, Nystroem KB (2003) Transformation of vegetable waste into value added products: (A) the upgrading concept; (B) practical implementations. Bioresource Technol 87:167–198
18. Masche M, Puig-Arnavat M, Jensen PA, Holm JK, Clausen S, Ahrenfeldt J, Henriksen UB (2019) From wood chips to pellets to milled pellets: the mechanical processing pathway of Austrian pine and European beech. Powder Technol 350:134–145
19. Matos AT (2005) Tratamento de resíduos agroindustriais (Treatment of agroindustrial residues). Universidade Federal de Viçosa, Viçosa, Brazil
20. Nussbaumer T (2003) Combustion and co-combustion of biomass: fundamentals, technologies, and primary measures for emission reduction. Energy Fuels 17:1510–1521
21. Rauter AP, Xavier NM, Lucas SD, Santos M (2010) Zeolites and other silicon-based promoters in carbohydrate chemistry. In: Horton D (ed) Advances in carbohydrate chemistry and biochemistry. Academic Press, Amsterdan
22. Sater O, De Souza ND, De Oliveira EAG, Elias TF, Tavares R (2011) Estudo comparativo da carbonização de resíduos agrícolas e florestais visando à substituição da lenha no processo de secagem de grãos de café (Comparative study of carbonization of agricultural and forest residues aiming at replacing firewood in the drying process of coffee beans). Revista Ceres 58:717–722
23. Sarkar N, Ghosh SK, Bannerjee S, Aikat K (2012) Bioethanol production from agricultural wastes: an overview. Renew Energy 37:19–27
24. Schenini PC (2011) Gerenciamento de resíduos da agroindústria (Management of residues from agroindustry). II. In: International symposium on management of agricultural and agroindustrial waste—v. 1 (lectures). Foz do Iguaçu, Brazil
25. The International Lignin Institute (2020) About lignin. http://www.ili-lignin.com/aboutlignin.php. Accessed Apr 2020
26. Tripathi N, Hills CD, Singh RS, Atkinson CJ (2019) Biomass waste utilisation in low-carbon products: harnessing a major potential resource. npj Climate Atmos Sci 2:35

27. United States Department of Energy (2004) Top value added chemicals from biomass—v. 1: results of screening for potential candidates from sugars and synthesis gas. DOE, Springfield
28. United States Department of Energy (2007). Top value added chemicals from biomass—v. 2: results of screening for potential candidates from biorefinery lignin. DOE, Springfield
29. Whittaker C, Shield I (2016) Short rotation woody energy crop supply chains. In: Holm-Nielsen JB, Augustine E (eds) Biomass supply chains for bioenergy and biorefining. Elsevier, Amsterdan
30. Vaz S Jr (2018) Analytical chemistry applied to emerging pollutants. Springer Nature, Cham
31. Virmond E, Rocha JD, Moreira RFPM, José HJ (2013) Valorization of agroindustrial solid residues and residues from biofuel production chains by thermochemical conversion: a review, citing Brazil as a case study. Brazilian J Chem Eng 30:197–229
32. Weinand JM, Mckenna R, Karner K, Braun L, Herbes C (2019) Assessing the potential contributions of excess heat from biogas plants towards decarbonising residential heating. J Cleaner Prod 238:117756
33. Zakzeski J, Bruijnincx PCA, Jongerius AL, Weckhuysen BM (2010) The catalytic valorization of lignin for the production of renewable chemicals. Chem Rev 110:3552–3599
34. Zaied M, Bellakhal N (2009) Electrocoagulation treatment of black liquor from paper industry. J Hazardous Mater 163:995–1000
35. Yun H, Clift R, Bi X (2020) Process simulation, techno-economic evaluation and market analysis of supply chains for torrefied wood pellets from British Columbia: impacts on plant configuration and distance to market. Renew Sustain Energy Rev 127:109745

6 General Remarks and Conclusions

Abstract

This chapter deals with the more relevant information described by the previous chapters in order to highlight the practical application of them.

6.1 Remarks

These remarks comprise the main points explored in each chapter in order to highlight the content of each ones.

We saw in Chap. 1 that the biomass production and uses come from the first human activities to survive the inhospitable environment. With the agriculture development, its production reached out huge quantities, generating wealth associated with environmental issues, which demands scientific knowledge and technologies of control and treatment to reduce negative impacts by means the modern sustainable vision. Nowadays, the world crop production is in the order of 7.26 Gtonnes of total production with a generation of 140 Gtonnes of dry biomass waste. This huge amount of residues creates an environmental problem that demands strategies and technologies for their treatment in order to promote economic value and social development, reducing negative impacts on the environment. On the other hand and according the World Bank, agricultural development is one of the most powerful tools to end extreme poverty, boost shared prosperity and feed a projected 9.7 billion people by 2050, accounting for one-third of global gross-domestic product (GDP) [9].

We saw in Chap. 2 that sustainability is a theme that call very attention in the twenty-first century by the global society. Nevertheless, it is a complex concept that looks to the future of our resources and life quality by means innovative business

S. Vaz Jr., *Treatment of Agroindustrial Biomass Residues*,
https://doi.org/10.1007/978-3-030-58850-2_6

strategies, take into account economic, societal and environmental impacts to be evaluated by metrics, as E-factor and life cycle assessment. Then, sustainability looks to the future of our resources and life quality by means smart strategies, which involves the 17 Sustainable Development Goals (SDGs) [8]. To achieve these goals for reduced environmental footprint and increased societal value, is paramount the investment in research & development & innovation to create new "green" technologies, especially for production systems and for transformation processes. Furthermore, the concept of circular economy and its application strategy can contribute to a holistic vision of the economic context of the biomass residues harnessing. The [2] defines it as *"a model of production and consumption, which involves sharing, leasing, reusing, repairing, refurbishing and recycling existing materials and products as long as possible. In this way, the life cycle of products is extended. In practice, it implies reducing waste to a minimum. When a product reaches the end of its life, its materials are kept within the economy wherever possible. These can be productively used again and again, thereby creating further value."* On the other hand, the life cycle assessment (LCA) is a very useful tool for the sustainability determination of industrial processes. It involves the evaluation of products and processes within defined domains, e.g., cradle-to-gate, cradle-to-grave and gate-to-gate, on the basis of quantifiable environmental impact indicators, such as energy usage, greenhouse gas emissions, ozone depletion, acidification, eutrophication, smog formation, and ecotoxicity, in addition to waste generated [4, 5].

We saw in Chap. 3 that chemical elements are renewed in the environment, being removed and returned to nature continuously by means biological, chemical and geological processes, constituting the biogeochemical cycles. The carbon footprint, a derived application of this cycle, became a strategy to understand and to reduce the generation of CO_2 from agroindustrial activities—considering that it is a greenhouse gas. Organic matter originates from the decomposition of residues from plant biomass and animal remains that, through chemical, physical and biological processes, undergo structural modification giving rise to a series of organic compounds, whose main representatives are humic substances. Additionally, the reduction of carbon dioxide emission from agricultural activities is an issue related to warming and climate changes because agricultural systems emit this gas from the land-use [1, 6]. Then, a low-carbon agriculture is a requirement for the 21th century.

We saw in the Chap. 4 that we can observe a concern from society with the negative impacts to environment and public health from agriculture and agroindustry. It is becoming paramount the development of a more environmentally and healthy friendly agribusiness. To achieve it are necessary technologies of monitoring—as real-time monitoring and probes—and technologies of treatment—as chemical, biochemical, thermochemical and physical processes for the generated residues. This Chapter deals with the introduction of more relevant technologies of monitoring and treatment to be applied to biomass residues. Furthermore, the proposal of a renewable carbon based-technologies for industrial purposes was presented. Moreover, [7] observed that agricultural-based industries produce vast amount of residues every year, and the majority of the agro-industrial wastes are untreated and underutilized, being disposed either by burning, dumping or unplanned landfilling. Furthermore, the proposal of a renewable carbon based-technologies for agroindustrial purposes is presented allied to adding value, biorefinery and bioeconomy concepts.

Finally, we saw in the Chap. 5 that agricultural biomass wastes are predominantly crop stalks, leaves, roots, fruit peels and seed/nut shells that are normally discarded or burned. However, there are some challenges in trying to determine the extent of crop-produced biomass in relation to what is a 'loss' (from production, post harvesting and processing), or a 'waste' (retail or consumer loss). Then, the choice of a treatment strategy is not a simple task; in some cases, will be necessary an association of processes—chemical, biochemical, thermochemical and physical categories—to achieve the better result. Moreover, aspects of environmental management and circular economy were discussed. The certification in ISO 14001 norm, Environmental Management System [3] help in the feasibility studies of monitoring and improvements in solid waste treatment projects and in the company in general. This norm provides a guideline to keep the organization within the laws regarding the company's field of action, offering an efficient environmental management system and, consequently, the solid waste management.

6.2 Conclusions

These conclusions were compiled from the content presented and discussed in each previous chapter, in order to reinforce to the reader those main points and statements in each ones.

Firstly, agricultural development is one of the most powerful tools to end extreme poverty, boost shared prosperity and feed a projected 9.7 billion people by 2050, and the world produces 7.26 Gtonnes of it for agroindustrial processing. However, this same biomass production generates 140 Gtonnes of waste with a heterogeneous chemical composition in different physical states that need the best technical and economic approaches to reduce their impact on the environment. This huge amount of residues can generates opportunities for its use as renewable industrial feedstock, according green chemistry and bioeconomy concepts, in a close relationship with the UN Sustainable Development Goals.

Despite sustainability be a theme that call very attention in the twenty-first century by the global society, it is a complex concept. However, sustainability looks to the future of our resources and life quality by means smart strategies, take into account economic, social and environmental impacts. To achieve the goals for reduced environmental footprint and increased societal value, is paramount the investment in research & development & innovation to create new "green" technologies, especially for production systems and for transformation processes. Metrics developed to evaluate these impacts are E-factor and LCA, which can be useful and complementary tools for the agroindustry. For instance, LCA can be applied to food waste to generate energy as an example of treatment of biomass residues according the circular economy approach. Then, in a practical way sustainability can contribute to more environmental, economic and social friendly processes and products to be deliver to the society.

Environmental chemistry, as a branch of Chemistry and Environmental Sciences, can contribute with the support of knowledge regarding to the biogeochemical cycles of the chemical elements in the nature, mainly carbon, as well as the genesis and fate of organic matter, specially that obtained from agroindustrial biomass. Moreover, a reduction on CO_2 emission and the water management will guarantee sustainable strategies based on scientific statements.

Nowadays, we can observe a concern from society with the negative impacts to environment and public health from agriculture and agroindustry. To minimize risks and negative impacts from biomass residues are necessary technologies of monitoring—as real-time monitoring and probes—and technologies of treatment—as chemical, biochemical, thermochemical and physical processes. The association between monitoring and treatment technologies promotes a reduction in the pollution potential of biomass residues and the valorisation of them. Additionally, a renewable carbon from biomass source as raw material can auxiliary in order to achieve a more sustainable agroindustry. Biomass residues from agroindustrial crops (e.g., soybean and sugarcane) are especially able to add value to their products chain. Biorefinaries are the best technological approach to achieve this statement in order to promote sustainable products and processes allied to profits. Moreover, it can be associated to green chemistry principles to increase this sustainability demand from the modern society.

Agricultural biomass wastes are predominantly crop stalks, leaves, roots, fruit peels and seed/nut shells that are normally discarded or burned but are in practice a potential valuable supply of raw material. However, there are some challenges in trying to determine the extent of crop-produced biomass in relation to what is a 'loss' (from production, post harvesting and processing), or a 'waste' (retail or consumer loss). Then, the choice of a treatment strategy is not a simple task; in some cases will be necessary an association of processes—chemical, biochemical, thermochemical and physical categories—to achieve the better result. For instance, the physical processing associated to the thermochemical processing. The application of treatment strategies using cogeneration (or combustion) for sugarcane bagasse, fermentation for pig slurry, coagulation and acid precipitation for black liquor, advanced on-site composting for brewing waste, pelletizing for wood chips, and subcritical water extraction for orange peel are very good examples of available technologies and scientific knowledge for the agroindustry to turns it sustainable. Furthermore, these technologies and knowledge are better explored when used within an environmental management vision.

Then, we can conclude that despite biomass residues from agroindustrial resources have a potential for environmental pollution they can be a source of renewable raw materials in order to promote the sustainability of agriculture and agroindustry.

References

1. Anderson TR, Hawkins E, Jones PD (2016) CO_2, the greenhouse effect and global warming: from the pioneering work of Arrhenius and Callendar to today's Earth system models. Endeavour 40:178–187
2. European Parliament (2018) Circular economy: definition, importance and benefits. https://www.europarl.europa.eu/news/en/headlines/economy/20151201STO05603/circular-economy-definition-importance-and-benefits. Accessed April 2020
3. International Organization for Standardization (2015) ISO 14001:2015 Environmental management systems. ISO, Geneva
4. Jiménez-González C, Curzons AD, Constable DJC, Cunningham VL (2004) Cradle-to-gate life cycle inventory and assessment of pharmaceutical compounds. Int J Life Cycle Assess 9:114–121
5. Monteiro JGM-S, Araujo OdeQF, de Medeiros JL (2009) Sustainability metrics for eco-technologies assessment, part I: preliminary screening. Clean Technol Environ Policy 11:209–214
6. Ruane AC, Phillips MM, Rosenzweig C (2018) Climate shifts within major agricultural seasons for +1.5 and +2.0 °C worlds: HAPPI projections and AgMIP modeling scenarios. Agric Forest Meteorol 259:329–344
7. Sadh PK, Duhan S, Duhan JS (2018) Agro-industrial wastes and their utilization using solid state fermentation: a review. Bioresources Bioprocessing 5:1. https://doi.org/10.1186/s40643-017-0187-z
8. United Nations (2020) Sustainable development knowledge platform. https://sustainabledevelopment.un.org/. Accessed April 2020
9. World Bank (2020) Agriculture and food home. https://www.worldbank.org/en/topic/agriculture/overview#1. Accessed April 2020

Printed in the United States
by Baker & Taylor Publisher Services